THE ENERGY CHALLENGE

Finding solutions to the problems of global warming and future energy supply

Geoffrey Haggis

Matador
9 De Montfort Mews, Leicester LE1 7FW, UK
Tel: (+44) 116 255 9311 / 9312
Email: books@troubador.co.uk
Web: www.troubador.co.uk/matador

ISBN 10: 1-904744-66-4

Typeset in 12pt Bembo by Troubador Publishing Ltd, Leicester, UK
Printed in the UK by The Cromwell Press Ltd, Trowbridge, Wilts, UK

 Matador is an imprint of Troubador Publishing Ltd

Our future is greater than our past.
So far we have mostly misapplied
The powers of the mind.
We have under-applied
The wonders of the human spirit.
The mind that created pyramids,
Warfare, great art, and science,
Has not yet reached maturity.
Everything we have done till now
Merely suggests the power of the human
Mind in its infancy.

...from Ben Okri's millennium address, delivered at the conclusion of Ways with Words, Dartington Hall, in the summer of '99, published as Mental Flight *(Phoenix House, 1999).*

ABOUT THE AUTHOR

Geoffrey Haggis won a mathematics scholarship to Trinity College, Cambridge and studied there for an honours degree in mathematics and physics. His PhD, at London University, was a study of the structure of water within the living cell. Later, by now a senior lecturer in the Physiology Department at the Edinburgh Medical School, he wrote with others, an *Introduction to Molecular Biology*. After 12 years at Edinburgh, following up an interest in electron microscopy, he worked for 20 years in agricultural research, at the Experimental Farm in Ottawa, Canada. He has spent periods of research and teaching at the University of California at Berkeley, the Karolinska Institute in Stockholm, the Erasmus University in Rotterdam, Paris University V and research centres in Venezuela and Brazil. Since retirement to Devon he has studied environmental issues, stimulated by the variety of lecturers of international standing who come down to teach at the Schumacher College in Dartington.

CONTENTS

Renewable energy sources within and around the UK are potentially
sufficient to supply all the electrical power which is generated today from
coal, natural gas and nuclear power stations. But this represents only one
fifth of the total energy used at present for all purposes. Much of the total
energy used today goes to heating buildings in winter and transport of all kinds.

Well designed houses require little heating in winter in the UK, just occasional
use of a wood-burning stove, or a heat pump, during long periods of grey weather.

Hydrogen is seen as the fuel of the future, for cars and other transport. Hydrogen
can be generated from natural gas, and if the carbon dioxide produced in this
conversion were returned to the gas fields, this would solve the global warming
problem, at least, till the gas runs out. Hydrogen can be produced from electricity
(by electrolysis of water) and can thus be derived from the renewable sources of
wind, tide, wave and sunlight. Hydrogen can also be generated from wood biomass.

To continue to rely on nuclear power for electricity generation would be very
expensive and carries too much risk. Coal could still be used in the future, with
sequestration of carbon dioxide under the North Sea. Domestic waste can be incinerated
to generate electricity, but this can make only a small contribution to energy needs.

CHAPTER 5

The volume of transport in the UK, and the energy this consumes, could be
significantly reduced in the future without any loss in the quality of life. In fact
there could be gain from greater dependence on local food production and reduced
road use.

CHAPTER 6

The growth of international trade is on collision course with the need to reduce
carbon dioxide emissions. Much of today's international trade benefits the rich,
but not the poor, and the terms of international trade need to be drastically revised.
Future negotiations to limit global warming could lead to some transfer of wealth
from rich to poor countries.

ILLUSTRATIONS & BOXES

ACKNOWLEDGEMENTS

I am, primarily, grateful to a number of people sending information in response to a phone call or email: Peter Fraenkel at Marine Current Turbines in relation to tidal energy, Peter Edwards at the Delabole wind farm in Cornwall for wind power, Peter Fardy at First Renewables in relation to the wood-fuelled power project in Yorkshire, Charles Clarke for the Holsworthy biodigestion unit, BNFL in relation to the decommissioning of the nuclear power station at Berkeley, GeoScience in Cornwall for heat pumps, Norsk Hydro for hydrogen generation from wind power, David Gee for the generation of hydrogen from methane, Jules Pretty at the University of Essex for copies of papers on sustainable agriculture, Shell for an update on their scenarios, Innogy for energy storage, Ventaxia for heat exchangers, BedZED for their housing project at Sutton, B9 for detail of their wood-chip CHP unit, National Wind Power for the wind farm at North Hoyle, the UN library in London and the ETSU library for sending reports, Sam Holloway at the British Geological Survey in relation to carbon dioxide sequestration under the North Sea, David Danbury at Bow Maurice Consultants in relation to yields from sustainable forest management.

Beyond this, I am grateful to my publishers for their vital input, to my friends: Carole Powell, Bruce Britton, Chris Marsh, Malcolm Baldwin, Robert Vint and Shirley Prendergast, who have read the typescript and made important contributions. I am, further, most grateful to those 'experts' whom I have persuaded to read it in part or in whole: Frijof Capra and Lester Brown for comment on Chapter 1, Jonathon Porritt and Richard Douthwaite for overall assessment and suggestions for improvement (which have been followed up). I have tried, as far as is humanly possible, to ensure that the facts of today here presented are accurate, though projections for the future remain speculative.

I have visited most of the sites described in this book to see what is happening on the ground. Articles in the press, in commercial brochures and in journals often describe hopes rather than current achievement. I am grateful to Dave Cowler for showing me round the decommissioned nuclear power station at Berkeley, to Stephen Wright for a view of one of the houses that Gusto were building at Millennium Green, to Graham Johnson for a tour of the Holsworthy biogas plant, at that time almost completed, to Martin Jolley for a look round the Welsh Biofuels plant, to Sam Watmore for a walk around the sustainably managed woodlands at Grascott Farm in North Devon, to David Hanstock and Andy Brown, of Progressive Energy, for discussion of coal gasification and to Aaron Custance for instruction on converting vegetable oil to biodiesel oil.

Finally, appreciation and thanks to Schumacher College, Dartington, for the opportunity to attend evening talks there over the past 10 years, notably, in relation to the material in this book, those given by: James Lovelock, Wolfgang Sachs, Amory Lovins, Vandana Shiva, Martin Kohr, Helena Norberg-Hodge, Paul Hawken, Jules Pretty, Ann Pettifor and Richard Douthwaite. Also thanks to Dartington Hall for the opportunity to attend weekend courses given by: Jonathon Porritt, Manfred Max Neef and Karl Henrik Robert.

I have, of course, to take responsibility for any failure to adequately represent all the thoughts about the future presented in these talks.

INTRODUCTION
THE CHALLENGE DEFINED

Many of the major changes of the past century: the increase in road and air transport, the construction of motorways and airports, the growth of global trade, have been made possible by the abundance of fossil fuels (coal, oil and natural gas). It became clear how much daily life in Britain depends on petrol and diesel oil, in the autumn of 2000, when a small group of truckers and farmers briefly cut off supplies to the forecourt pumps. Without petrol and diesel distribution, people could not drive to work and children could not be driven to school. Soon, it seemed, trains would stop running and supermarket shelves would be empty.

This dependence on the fossil fuels is not seen only in Britain. The global economy today is based on transport: holiday travel, transport of food and manufactured goods over long distances, logs trucked out of the rain forests, all on the basis of abundance of oil. However, it has become increasingly clear, during the early years of the new century, that the development of a global economy so dependent on the fossil fuels for growth and prosperity is setting the human race on a course which may prove disastrous. In the first place, the level of carbon dioxide in the atmosphere is rising towards a point at which the resulting warming of the Earth will destabilise the global climate. In the second place, reserves of fossil fuels are finite. Many analysts now believe that global oil production will peak before 2020, and that the peak in natural gas production will follow less than 20 years later.

Renewable energy sources must be developed as rapidly as possible and in many areas there can be reduction in energy use. For example, demand for home heating can be dramatically reduced by good insulation. Coal, which will remain abundant for some time, can perhaps still be used in a nonpolluting way and, in the last resort, it may be necessary to accept the risks and hazards of nuclear power. These are the issues to be explored in this book.

There is still doubt in some people's minds as to whether global warming is actually happening and, if it is, whether this is just part of a natural cycle. So what is the evidence for this rather dramatic suggestion that human activity can influence climate on a global scale? There can be no doubt that the atmospheric concentration of carbon dioxide has been increasing steadily over the past century, and more steeply since about 1960. Atmospheric concentrations of carbon dioxide are measured at Mauna Loa observatory in Hawaii, far from any source of industrial pollution, and the level at the time of writing ('04–'05) is around 375 parts per million (ppm), 34% above the preindustrial level of 280 ppm. It is possible to estimate with reasonable accuracy how much carbon dioxide is released each year into the atmosphere from the burning of fossil fuels, and roughly how much is released by the destruction of forests. The total amount is huge, more than sufficient to account for the observed rise in atmospheric concentration. In fact the rise would be twice as large if roughly half of this emission were not absorbed by the oceans and by forest areas where there is net growth. The growth of forests takes carbon dioxide from the atmosphere with the carbon stored in the wood of the trees. The destruction of forests releases this stored carbon as carbon dioxide.

> The pre-industrial level of carbon dioxide is measured in air trapped in Arctic ice layers laid down in the mid 18th century.

As long ago as 1827, Fourier appreciated that the earth's atmosphere acted 'like the glass of a hothouse' trapping the heat of the sun, and the first calculations of how much an increase in the level of atmospheric carbon dioxide might contribute to global warming were made by Arrhenius in 1896. He found (in close agreement with the upper range of today's predictions) that a doubling of the atmospheric carbon dioxide above the preindustrial level might warm the earth by 5 to 6°C [1].

> After the break-up of his marriage to a beautiful research assistant, Arrhenius (turning to sums for solace) spent over a year in self-imposed isolation, making calculations which would take modern computers only seconds [1].

There is thus very good reason to expect some global warming as a result of today's increasing level of atmospheric carbon dioxide. In recent years, the evidence that this warming is actually happening has become stronger, from

When Sir Edmund Hillary visited Everest in 2004 he noted that, while back in 1953 snow and ice reached all the way down to base camp, now it ends five kilometres above this point [6].

direct measurement of temperature, the observed retreat of glaciers, thinning of ice at the North Pole, etc. In July '03, the Geneva-based World Meteorological Organisation, issued a press release which highlighted extremes in weather and climate occurring all over the world [2]. In Britain, and more generally over Western Europe, temperatures rose to new record levels.

Today, in the early years of the new century, it is perhaps just possible to argue that the warming observed, and the increased incidence of extreme weather conditions, could be due to a natural climate variation, in no way connected with the vast amount of carbon dioxide which the human race now puts into the atmosphere each year, but this is becoming an increasingly untenable position.

Climatologists recommend that global carbon dioxide emissions should be reduced well below 1990 levels by 2050 to avoid the possibility of quite catastrophic consequences: rising sea levels, more violent storms, serious flooding in some areas, extended drought in others.

Water vapour is also a greenhouse gas.

Aside from carbon dioxide, other gases: methane, nitrous oxide, fluorocarbons, etc., also act as greenhouse gases, trapping sunlight. The atmospheric level of some of these other gases is also rising. In a report published in 1990 [3] the Intergovernmental Panel on Climate Change estimated the warming effect of each of these gases and also included climate modelling computer studies of the sea level rise and climate change to be expected over the coming century if we continued 'business as usual', that is, continued to burn fossil fuels at the rate they are being burnt today. Among the scenarios considered in this report is one in which carbon dioxide emissions are allowed to rise by 2% per year until 2010 and then reduced by 2% per year thereafter, stabilizing the atmospheric level at about 450 ppm. This is essentially the scenario chosen for the curve on p.4.

Predictions of climate change based on computer modelling naturally involve a measure of uncertainty. They are limited in their accuracy, amongst other things, by the difficulty of predicting ocean current changes and determining the extent to which very fine air-pollution particles modify the reflectivity of clouds.

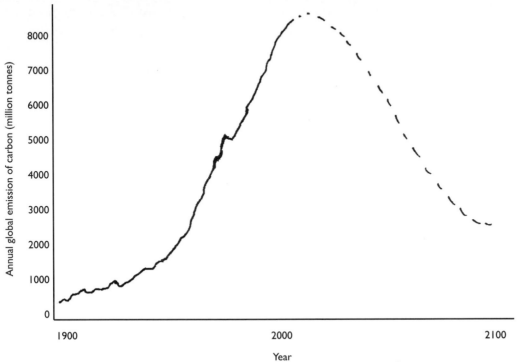

THE RISE AND PROJECTED FALL OF CARBON EMISSIONS

A curve showing how the global consumption of fossil fuels grew during the 20th century (full line) and how it would have to be reduced in the 21st century to stabilise the atmospheric carbon dioxide level at 450 parts per million, that is, at 60% above the pre-industrial level of 280 parts per million.

(Source: *Contraction and Convergence*, Aubrey Meyer, Green Books, 2000)

However, since 1990, not only has the evidence for global warming become stronger, there is also greater concern about possible rapid, irreversible changes in climate which might result from continuing 'business as usual'. The real worry is the possibility of 'positive feedback'. For example, forests cannot adapt to too rapid a temperature change. It takes time for new tree species, adapted to warmer conditions, to replace the existing forest. If, at a certain point, the trees of the Amazon begin to die, rot and release carbon dioxide into the atmosphere, this could spur further warming to produce a rapid, runaway warming effect.

Computer modelling of future climate change cannot yet define an upper limit of atmospheric carbon dioxide which it would be dangerous to exceed. There seems to be a rough consensus among climatologists that doubling the pre-industrial level would be risky. The downward fall of the curve on p.4 shows the emission reduction needed to stabilize the atmospheric level of carbon dioxide at

As the glaciers retreat on high mountains and as the snow line moves North in Arctic regions, snow and ice, which reflect sunlight, are replaced by bare rock and earth which absorb the sun's energy. This also leads to positive feedback. This effect is exacerbated if methane and carbon dioxide are released from the warming tundra.

THE 'NEEDLE' OF FOSSIL FUEL USE

The curve of p.4 on a timescale of two millenia, rather than two centuries. The period of high fossil fuel consumption is seen as just a blip in the history of the earth.

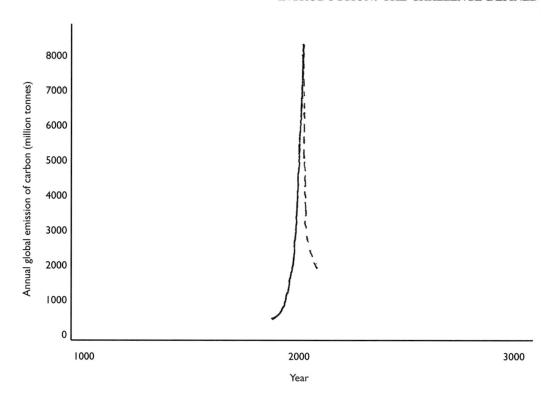

a 'safer' 60% above the pre-industrial level. Only time will tell what is a truly safe level for human survival. For the Maldive Islanders the safe level is already passed.

Can this reduction in carbon dioxide emissions be achieved? It seems that to make such drastic change in our lifestyle, reducing the use of fossil fuels each year, is going to be very difficult. A conference in Kyoto, in late '97, sought to establish international agreements, to at least begin to limit carbon dioxide emissions, but ratification and implementation of these agreements is proving difficult because this will involve, essentially, a total turnaround for the global economy, a change probably from growth of global trade to reduction in global trade, since transport involves heavy energy use. It is not surprising that economists and politicians have great difficulty in facing up to the need for such fundamental realignment of the economy.

In this book I explore the changes that will be needed to move from an economy so dependent on fossil fuels to one which, ultimately, will function on renewable energy alone. Over the next 100 years, the development of renewable energy sources will be spurred at first by concern about climate change. However, global oil production is expected to peak sometime between 2010 and 2030. Once this point is reached, oil supply will no longer meet demand and the price of both oil and natural gas will rise, unevenly but inexorably. The reduced availability of fossil fuels will provide a further spur to renewable energy development, as the century progresses.

The UK faces an immediate challenge. By 2020, there will be little oil or natural gas left under the North Sea and all but one of today's nuclear power stations will have been closed down. Unless the right decisions are made now, Britain could become critically dependent, for both electricity generation and central heating, on gas coming from politically unstable regions thousands of miles away.

Any number of disasters are possible before 2100: earth's collision with an asteroid, nuclear war through computer failure, the Gulf Stream ceasing to flow putting Britain and Northern Europe into a mini ice zone, major volcanic eruptions, nuclear accidents on the scale of Chernobyl, etc. But it is also possible that we may have the wisdom to prevent, or minimise, these catastrophes.

The slowing or stopping of the Gulf Stream is one of the possible consequences of global warming [4].

By 2100, the human race, if it survives at all, will be living in a more sustainable way. This is simply a definition of the term 'sustainable'. For example, the only areas where fishing will still be continuing will be where fish stocks have been conserved. In other areas there will be no fish left in the sea. In farming, sustainability means eliminating soil erosion and ensuring the return of nutrients and organic matter to the soil, to maintain fertility from one generation to the next. The peak on p.5 shows the curve of p.4 on a scale of two millennia, rather than a scale of two centuries. The period during which fossil fuels are used so intensely is seen as a brief blip in the history of the earth.

The early chapters of this book are a study of energy problems and potential

UK = United Kingdom
US = United States
EU = European Union

solutions for the UK. This study can be conveniently made since the UK Department of Trade and Industry (DTI) produces each year a detailed study of energy use during the previous year and has also published reports on the potential for offshore wind power around the UK, and for photovoltaic power, that is, the generation of electricity from sunlight. The problems and solutions for Britain are similar to those of continental Europe, though in the UK renewable resources are, as yet, less well developed than those of Germany, Spain and Denmark and houses are less well insulated than those in Sweden.

What we find, in this study, is that it is going to be difficult to generate from renewable sources the amount of energy used today in the UK, but that the present lifestyle is very wasteful of energy. Considerable energy saving is possible without any loss of comfort or decline in the quality of life. In fact, the quality of life may be enhanced by reduced energy use.

The outline for the book is indicated in the summary for each chapter on pp. v–vi. The first five chapters build a scenario for the UK in 2050, summarised in the 'box' on pp. 98–99. The Royal Commission on Environmental Pollution and the Prime Minister's Performance and Innovation Unit have produced a variety of energy supply scenarios for 2020 and 2050 [5]. This is a useful first step, but eventually a future scenario has to be chosen, at least provisionally, and a strategy developed to meet the energy challenge. Global oil production is expected to peak before 2020, and once oil production peaks the price of both oil and natural gas will rise. Global production of natural gas will also have peaked before 2050. The 21st century will, in a sense, echo the 20th with energy crises replacing major wars. Peak oil can be expected about 2014 and peak gas about 2039. Of course, these dates are still uncertain, but I have assumed for my scenario that, by 2050, conventional oil and natural gas will have risen so much in price, and the consequences of global warming will have become so severe, that these fossil fuels will no longer be used. Coal will still be plentiful, in 2050, as will be the so-called 'unconventional' oil, in shales and tar sands, but their use will only be allowable, by that time, if the carbon in the coal and oil can be captured and stored, rather than released into the atmosphere. We must begin to reduce our dependence on the fossil fuels to control global warming and we shall also be forced to do so as global production of oil and gas peak and begin to decline.

It is important that the right decisions be made now, to prepare for the future, because global warming is gathering pace and we may be already close to a point of no return, a 'tipping point' at which positive feedbacks kick in and the human race becomes powerless to prevent catastrophic climate change.

It is vital for human survival that solutions be found to the problem of global warming, but there are two other challenges in the new century:

- feeding a global population increasing from around 6 billion now to around 9 billion by mid-century

- reducing the growing disparity between rich and poor and distributing more fairly the resources of the earth.

In later chapters of the book I explore the ways in which the need to adapt to renewable energy sources could favour sustainable farming and a fairer global distribution of wealth. Those countries which are poor today but have abundant sunshine will be rich in energy. Countries with less sunshine have more potential wind, hydro, wave and tidal power. If the change to renewable energy can be achieved, this would avoid the political tension and potential conflict which might arise from competition for the last oil and natural gas reserves remaining in Iraq or Saudi Arabia. There would no longer be the need for pipelines through rain forests or exploration for remaining pockets of oil and gas in the still pristine areas of the world.

The Notes at the end of the book (referenced in the text by numbers in square brackets, [1] etc.) are sometimes source references but, for the most part, give further detail, for the reader who wishes to explore the issue more deeply. Where the term 'Britain' is used (to break the monotony of using 'UK' all the time) this is to be read as 'Britain and Northern Ireland'.

Energy calculations throughout this book are based on data for the year 2000. 'Today' and 'at the time of writing' should be read as 'in the early years of the 21st century'.

Inevitably, some of the forward projections in this book will prove wrong. The one certain thing about the future is that it will bring surprises and confound predictions. Nevertheless, it is important that we begin to develop some long-term aims, while tackling the immediate problems which seem so overwhelming at the start of the new century. I have tried to include sufficient background detail, in text, 'boxes' and Notes to allow the reader to question my projections at every point, and develop her/his own vision for the future.

CHAPTER I

RENEWABLE ENERGY: WILL THERE BE ENOUGH?

Renewable energy is everywhere, in the sun and the wind and the waves, but nowhere, except at the foot of a large dam, is it available in the concentrated form in which we find the energy of coal and oil. This makes renewable energy eminently suited to the needs of today's poor countries, where energy consumption per capita is low. Electricity for a water pump, a water purification unit, or for lighting, can make a significant difference to life in an African village, while requiring little power. By contrast, we shall see that renewable energy sources are inadequate to maintain the energy-wasteful lifestyle which prevails today in the industrial world. In this first chapter, I want to look at some of the main sources of renewable energy available in, and around, the UK and put some figures into the broad equation of future need. How much energy can sun, wind, wave and tide supply? This is essential background for the discussion of later chapters.

Wind power

There are wind farms in many areas of Britain now, particularly in Cornwall, Wales and Cumbria. Modern turbines are very robust and reliable, designed to stop turning at winds above 50–60 mph (80–100 km per hour) and turn their blades out of the wind, to survive a gale. The generators, in the boxes behind the spinning blades, feed 50 cycle/sec alternating current directly into the National Grid. Electricity is thus generated in a totally clean way from the wind, a resource that will last for ever, and cattle can graze, undisturbed, below the turbines.

We shall see, in later chapters, that wind power is going to be one of the main

**A WIND FARM OF
ABOUT 30 TURBINES**

(Copyright: Woodfall
Wild Images)

energy sources for the future, not only to generate electricity for direct use but also electricity used to produce hydrogen. This seems the ideal answer for clean, sustainable, energy generation, but there is still debate about installation of more wind farms in the rural areas and moorlands of Britain. Beautiful as these turbines are in themselves, necessary as they may be for energy supply in the future, they do not really fit into the traditional picture of the countryside. There is much uncertainty today about the future of the British countryside and the future for farming. I am going to suggest, in a later chapter, that farming in Britain will be of central importance in a 'renewable energy world' with reduced dependence on imported food. There may be further movement towards organic, or near organic, farming with restoration of wild flowers and song birds to 1950 levels. Be that as it may, the countryside is, and hopefully will remain, a valuable asset for tourism and for escape from the stress of city life. The windy areas of Britain

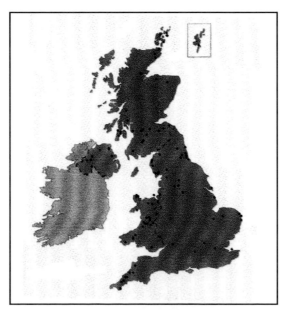

**WIND FARMS IN
THE UK (2005)**

(Source: British Wind
Energy Association)

are often the most beautiful wild areas for walking and recreation. If all the high ground were covered with wind farms an important element of tranquillity would be lost. Renewable energy sources must be developed today as rapidly and extensively as possible but ultimately a balance must be struck between the need for clean energy and the need for biodiversity, wild life and for areas of wilderness and tranquillity.

Perhaps the real potential for onshore wind power lies in small scale local development. An isolated turbine supplying power to a village, or a small cluster of turbines supplying a small town, can be sensitively fitted into the countryside. Combined with energy storage, to be discussed later, this can allow local self-sufficiency in energy supply. But where wind power is to make a significant contribution to the needs of a larger town, London for example, the main development has to be offshore. Winds, in any case, are steadier and stronger offshore.

In the scenario for 2050 in Chapter 5, I have allowed for 7000 onshore turbines, seven times the number installed by 2003.

The UK Government began to take global warming more seriously after the floods of late 2000. There was increasing realisation that this warming might not just lead to mild winters and balmy summers, but that Britain might be adversely affected by climate change. Computer model projections suggest that most of Britain, except the South East corner of England, could become wetter as well as warmer, with greater likelihood of violent weather events: stronger winds and short bursts of localised heavy rain which could cause flooding.

In a White Paper 'Our Energy Future' published in Feb '03 [1] the UK Government accepts the long-term need for a 60% cut in carbon dioxide emissions by 2050 while, for the time being, leaving undefined exactly how this is to be achieved. The immediate targets are to generate 10% of electrical power from renewable sources by 2010, and then 20% by 2020. If these targets are to be met, offshore wind power is going to have to make a large contribution.

In the first stage of this development, farms of 30 turbines each have been installed off the North Wales coast, at North Hoyle, five miles (8 km) out to sea, and on the Scroby Sands, two miles off the Norfolk coast at Great Yarmouth. It has been concluded, after five years of environmental study, that the latter project will not disturb the breeding patterns of bird and seal colonies on the sand bank, though clearly the effect on sea birds and marine life must be kept under review.

In a computer-generated view of the wind farm at North Hoyle, as seen from Rhyl Beach (prepared for planning application), the turbine masts can just be discerned on the horizon (picture p. 16). Environmental studies have shown that these turbines are outside the foraging range of local sea birds and that dolphins and porpoise are not likely to be disturbed by them. Whales, by contrast, are extremely sensitive to noise in the frequency range produced by wind turbines. The turbines installed off the North Wales coast and at Scroby Sands are rated at 2MW, each one supplying the electricity needs of 1500 homes [2]. Larger 3MW turbines are expected for future offshore development.

One Megawatt (MW) = 1000 kilowatts (kW)

Denmark, at one time, planned to install some 2000 turbines over the next ten to twenty years, mostly offshore. If these plans went ahead, this development would supply 50% of Denmark's electricity needs. In principle, the UK offshore programme could be extended to allow wind power to meet 50% of UK electricity needs. However, to supply this amount of power around 16,000 3MW turbines would be needed [3]. The population of the UK is about ten times that of Denmark so many more turbines would be needed, as compared with Denmark. Taking the UK coast line as roughly 3000 miles, a total of 16,000 turbines means a farm of 30 large turbines roughly every six miles around the coast.

Wind turbines on land are usually smaller than those offshore and winds on land are more variable. About 35,000 onshore 1.5 MW turbines would be needed to produce the same power as 16,000 3 MW turbines offshore. Roof-top turbines could make a significant contribution if there were millions of them [11].

Of course there could be fewer, larger farms, and this is exactly what is planned for the second stage of offshore development, in which larger farms are to be sited in three regions: the North West, from North Wales to the Solway Firth, the Greater Wash area off Norfolk and Lincolnshire, and the Thames estuary (see map on p.27). These farms will be within the territorial limit of 12 nautical miles (22 km) and in depths of less than 30 metres. Later it is expected that improved technology and changes in legislation, will allow turbines to be installed well

WIND FARM OFF RHYL BEACH, NORTH WALES

(Source: Npower Renewables)

offshore out of sight of land, in depths of up to 50 metres and beyond the territorial limit.

Provided these farms are carefully sited, so that they do not interfere with migrating birds flying into the estuaries, this seems an ideal solution. Peter Edwards, who set up the first land wind farm in Britain, at Delabole in Cornwall, tells a story of a flock of birds once streaming through between the turning blades of one of his turbines. When he went out later, expecting to find the ground below strewn with casualties, he found none. So birds may be able to avoid the danger posed by turbines.

Swans flying in from Siberia may not as readily avoid the blades as sparrows in Cornwall.

We can now ask what it will mean, if these offshore developments go ahead? To meet a target of 10% of electrical power from renewable sources, supposing, say, half this power is to come from offshore wind, will require the installation of 1600 3MW turbines (five turbines installed each week from Jan '04, if this target is to be achieved by 2010). A programme on this scale, continued through the

This illustration shows the size of a 2 MW turbine in relation to other land and sea features. The larger 3 MW turbines reach up 130 metres and the blades sweep out a circle 100 metres in diameter.

(Source:www.offshorewindfarms.co.uk)

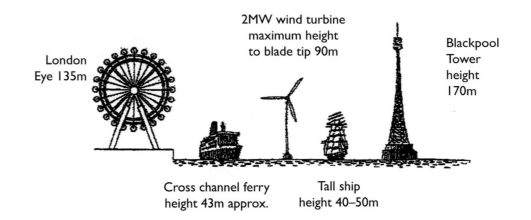

London Eye 135m

2MW wind turbine maximum height to blade tip 90m

Blackpool Tower height 170m

Cross channel ferry height 43m approx.

Tall ship height 40–50m

m = metres

coming century, could lead to the installation of 16,000 turbines by 2070, providing half of electricity supply, at today's level of consumption.

Now let us pause and take thought. Electricity consumption in Britain today is only about *one fifth* of *total* energy use (see pie chart on p.19). A lot of additional energy is used to heat homes and other buildings, for industry and for transport. The 16,000 turbines would only be supplying *one tenth* of UK total energy needs. There is huge potential for wind power development around the UK. In waters up to 30m deep and within the territorial limit, there is more than enough potential to supply all electricity needs. Going out to depths of 50m, and beyond the territorial limit, there is more than enough for all energy needs [4]. But to derive more than one tenth of the energy used today, from offshore wind, would require more than 16,000 turbines.

Modern wind turbines are very reliable but the wind, of course, is not. At the present time, when wind power makes only a small contribution to total energy supply, the intermittent nature of this source does not cause any problem. Its variability falls within the normal fluctuation of supply and demand. However, as renewable energy begins to play a larger role, energy collected on windy days needs to be stored, a point we return to later in this chapter. Wind power is favourably linked to the seasons, with more wind in the winter when more energy is needed. Energy from sunlight, which we shall be discussing later, is less favourable from this point of view, being more abundant in summer. However, energy from sunlight is favourably linked to the working day, providing energy during the daylight hours when it is needed for factories and offices.

Wind power is making an increasing contribution to energy needs in Europe. Germany and Spain, as well as Denmark, are well ahead of the UK in this respect. For the United States, installation of more wind farms could substantially reduce carbon dioxide emissions. There is enough potential wind power in just three states, North Dakota, Kansas and Texas, to meet all US electricity needs. Wind power has also an important role to play in meeting the energy needs of developing countries. There is an initial capital cost for installation of the turbines, but thereafter maintenance costs should be low.

In addition to wind power there are, of course, other renewable sources and it is widely accepted amongst environmentalists that a mix of sources will be needed in the future.

Fuel oil from seeds

We now consider a very different kind of renewable energy. It is possible to grow oilseed rape, or a comparable crop, and press the seeds to produce vegetable oil. This can be readily converted to 'biodiesel', a thinner oil which can directly replace normal diesel oil as a fuel for cars and lorries [5]. The use of a fuel of this kind is carbon neutral, that is, adds no carbon dioxide to the atmosphere, since the carbon dioxide emitted by the cars and lorries is balanced by the carbon dioxide taken from the atmosphere while the fuel crop is growing.

Research at the Folkecenter for Renewable Energy in Denmark [5] has shown that winter rape is an ideal crop for this purpose. It gives a high yield without use of pesticides. It can be beneficial in crop rotation, in suppressing weeds and

THE PROPORTION OF DIFFERENT FORMS OF ENERGY USED IN THE UK (for the year 2000)

For coal and natural gas this chart shows final energy use. It does not include the coal and gas used in the generation of electricity [3].

Electrical power stations in 2000 were:

29% coal fired
39% natural gas fired
24% nuclear

with 5% of electricity imported from France and 3% coming from renewable sources.

(Source: DTI Energy Statistics, 2000)

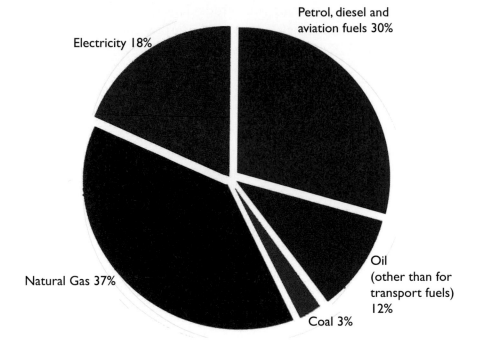

Petrol, diesel and aviation fuels 30%

Electricity 18%

Natural Gas 37%

Coal 3%

Oil (other than for transport fuels) 12%

I shall use the term 'biofuels' for liquid fuels derived from plants, such as seed oil and ethanol, and the term 'wood biomass' for wood chips from coppice or woodland.

building up organic matter in the soil. After the oil is pressed out, a seed cake remains as an animal feed.

Oil from 10% of the agricultural area in Denmark could fuel 25% of their road passenger traffic or all their agricultural machinery. For the UK, unfortunately, 10% of agricultural land (that is, about 1.8 million hectares) set to winter rape could supply only some 7.5% of the fuel used today in the UK for cars alone, not including other passenger and freight transport by road [6]. Also the production of biofuel crops is, in some measure, in competition with food crop production, even if winter rape is sown. Once again we are faced with the fact that this form of renewable energy is nowhere near sufficient to meet current needs.

Winter rape, and comparable crops, could play an important role in the future, integrated into sustainable farming and producing, locally, the fuel for tractors and farm machinery. But biofuels can go little further than this to meet future energy needs in the UK, if agricultural land is to be used mainly for food production.

In Chapter 5 a scenario is developed for life in Britain in 2050, based mainly on energy generated renewably in and around the British Isles. In this scenario farmers grow oilseed rape on part of their land, to fuel their tractors, farm machinery and small vans to take their produce to market, much as farmers in earlier times set land aside for grazing and for hay for their horses.

It is possible to grow sugar beet, as an alternative to oilseed crops, and ferment the sugar to ethanol for use as a fuel, but the same considerations apply: the area of beet required to fuel current transport needs is unrealistically large.

Possibly the yield from rape, or other oilseed crop, could be increased in the future through traditional breeding or genetic engineering, reducing the area of land needed to produce a given amount of oil. However, there is an ultimate limit to the energy a plant can store in its seeds, set by the limited efficiency of photosynthesis and the fact that a plant, in addition to producing seed, must also grow leaves to capture the energy of sunlight and roots to absorb water and nutrients.

Photovoltaics (PV)

A potentially much greater source of renewable energy lies in direct generation of electricity from sunlight ('photovoltaics', photons producing volts). We are all familiar now with the small hand-held calculators, which need no batteries but which work only when light is falling on the little PV panel above the number display.

A more significant amount of power can be produced from the larger PV panels (of area a little over half a square metre) which are now available to mount on the

Used cooking oil can be converted to biodiesel, but this cannot make a significant contribution to national oil demand (only 0.25%, see *Heat*, George Monbiot, Allen Lane, 2006).

SUSAN ROAF'S HOUSE AT OXFORD

W = water-heating panels

(Copyright: Nigel Francis)

sq.m. = square metres

roof, or south-facing facade, of a building. Susan Roaf's house at Oxford (above) has 48 of these panels, made by BP Solar, on the south-facing side of the roof, covering an area of 30 sq.m. She gains enough PV power from this area (on average over the year) for household electricity and also to run a small electric car for local journeys [7]. These PV panels should not be confused with simple steel box panels on a roof, designed to heat water from sunlight. Susan Roaf's roof has four flat panels for water heating in addition to the PV panels.

PV panels and tiles incorporate a thin layer of silicon, infiltrated with small amounts of boron and phosphorus, which allow the energy of sunlight (or daylight on a dull day) to generate an electrical voltage.

Susan Roaf has shown that a south-sloping roof area of 25–30 sq.m. is adequate to supply the electricity needs of an average home, if the home is fitted with energy-saving light bulbs and the latest energy-saving appliances (fridge, clothes washer, etc.). As noted above, Susan Roaf also runs a small car from 30 sq.m. How does this compare with average home consumption today?

The average home today consumes 4500 kWh/year [2]. Susan Roaf's PV roof produces 2800 kWh/year. Of this she uses 2000 kWh/year for domestic need and 800 kWh/year to run the little car [7]. This gives us some idea of how much electricty could be saved if all homes were equipped with the most efficient energy-saving lights and appliances. Domestic consumption could be halved. This is a most important point. Instead of planning, for example, for new nuclear power stations to meet today's electricity need, or expected increased need in the future, it would make more sense to be considering how best to foster wider use of energy-saving lights and appliances.

It is expected that, through large-scale production, or through development of cheaper production methods, the cost of PV panels and roofing material will come down significantly over the next ten to twenty years. A DTI report [8] estimates that if all the roofs and facades of buildings in the UK were covered with PV panels with an efficiency of 10% (the overall efficiency achieved by Susan Roaf's roof) this area could supply 50% of electricity needs. 100% of UK electricity needs could be met in this way if further research raised conversion efficiency to 20%. This increase in efficiency may be achieved, but probably at a cost. If the cost comes down, just for panels of moderate (10%) efficiency, this would allow the widespread installation of PV roofing in all new housing.

PV panels and roof tiles produce D.C. (that is, direct current) but this can be converted, by means of an 'inverter' to the usual 50c/s A.C. (alternating current) of the UK National Grid. With an inverter box installed, PV can feed electricity into the grid on sunny days. Electricity is drawn from the grid at other times. If the electricity supplier can be persuaded to buy and sell at parity (that is, if the price paid by the supply company for a unit of electricity equals the cost to the

INSTALLING PV TILES

A variety of PV panels and tiles are now available.

(Source: solarcentury.com)

consumer of a unit drawn from the grid) then the meter can run backwards when the sun shines and the consumer pays only for net electricity drawn.

This hope that PV could supply most of Britain's domestic electricity in the future involves not only the expectation of reduction in PV costs but also a further level of optimism, that is, that Britain continues to enjoy the average level of sunshine which has been the norm over the past 100 years. If climate change brings almost constant cloud and rain to the UK, as unfortunately it might, then this estimate would have to be revised.

The future for PV in developing counties is most exciting. Already, with the cost of PV still high, 200,000 solar home systems have been installed in China,

PV IN AFRICA

(Copyright: Shell International)

100,000 in Kenya, 70,000 in India, 50,000 in Indonesia, 50,000 in South Africa and 40,000 in Mexico. 3,000 PV systems for hospitals, community centres and schools have been set up in the Philippines [9]. These developments are being funded by the World Bank and other agencies. PV, combined with a heat pump (described in the next chapter) is ideal for air conditioning. It produces power at the exact times, and in exactly those places, that air conditioning is needed.

PV panels and roof tiles should prove very durable, since no moving parts are involved, and they require no maintenance beyond keeping the outer surface clean enough to allow light through to the sensitive silicon layer.

Tidal current turbines

The first UK experimental tidal turbine was installed off the North Devon coast at Lynmouth in '03. The support tower protrudes above the surface of the sea, but is less intrusive than a wind turbine (picture p.26). The blades turn so slowly that they should pose no danger to fish and, in fact, areas of tidal turbines, as well as offshore wind farms, could be a great help in restoring fish stocks, since these areas would be off-limits for trawlers.

In favourable situations under-water turbines could be mounted on the pillars supporting wind turbines.

It is estimated that tidal power in the Pentland Firth, between the Orkney Islands and mainland Scotland, could alone produce 10% of today's electricity supply and that a comparable amount could be generated from tides around Alderney in the Channel Islands (see map on p.27).

Wave power is another possibility as yet still only at the experimental stage. If harnessing wave power involves concrete channels along the coast, focussing the incoming waves to pressure-driven electricity generators (which is one possibility) this is not an environmentally satisfactory solution, if applied on a large scale.

It is also possible to create a lagoon, by building a surrounding sea wall in shallow tidal areas and generate power as the tide flows in and out of the lagoon [12].

Wherever there is flowing water there is the possibility of generating electricity. The monks at Buckfast Abbey, in Devon, installed two water-driven turbines in the bank of the river Dart as long ago as the 1950s. Together these supply 160 kW of power when the river is flowing well. Part of the river flow is diverted into a leat 8 feet wide and 8 feet deep (2.5 metres by 2.5 metres) with its entry some 200 metres up-river from the turbines.

This type of small-scale hydro-power is being installed successfully in developing countries, where there are rivers and streams in remote areas and where energy needs are modest. At the present time about 1% of electricity in Britain is generated from hydropower, mainly from dams in suitable locations.

TIDAL CURRENT TURBINES

(Source: Marine Current Turbines TM Ltd)

Storing renewable energy

One of the problems with wind, wave, PV and, to some extent, tidal current power is the intermittent nature of the energy supply. The turbines are not always spinning, nor the sun shining brightly, at times of peak demand. Electricity can be easily stored on a small scale, as in the rechargeable batteries of a flashlight or in a car battery but it is difficult to store electrical energy on a large scale. Without storage, backup generation must be available to meet demand over periods when power from wind, wave and PV falls away. A medium-scale storage system has been set up at Little Barford in Cambridgeshire (described in the 'box' on p.132) but this plant has been 'moth-balled' because such storage is not yet

POTENTIAL SITES FOR TIDAL CURRENT TURBINES AROUND THE UK [10]

(Copyright: Guardian Newspapers, 2003)

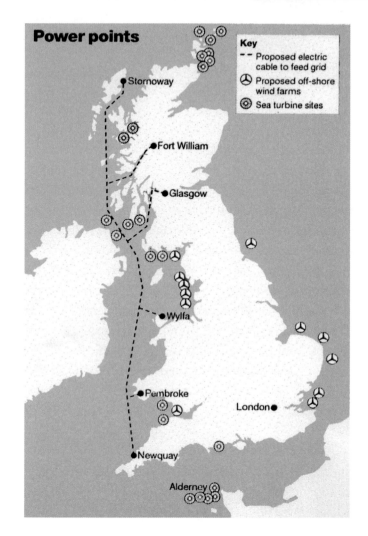

economically viable. At the present time, wind and PV contributions are small and the National Grid has the capacity to meet peak demand from fossil fuel and nuclear generation. However, if renewable sources are eventually to play a major role in electricity supply, it will be necessary to develop some form of storage. An installation of the kind set up at Little Barford could not store enough energy from summer PV for winter use, but it could store PV power generated on a sunny day in winter to cover early evening peak demand on that day, or energy generated by wind turbines at night, for use the following day. We shall see in Chapter 3, that electricity can be used to generate hydrogen and vice versa, so that electrical energy can be stored as hydrogen, but there are costs and energy loss in such conversions.

Leaving aside, for the moment, this rather important question of how energy from intermittent sources is to be matched to demand, or integrated into a hydrogen economy, it certainly seems that *enough* power could be generated from wind, wave, PV and tidal flow to allow the UK to meet electricity needs in the future from renewable sources, though even if PV costs come down, it will take some time for PV roofs to gradually cover new and existing buildings. In the scenario for the UK in 2050, developed in Chapter 5, it is supposed that half the electricity needed will come from wind power and half from PV, tide and wave. It is further supposed, for this scenario, that electricity consumption later in the century will remain much as today. The number of homes in the UK will increase, but consumption per home will be reduced, through use of low-energy light bulbs and energy-efficient appliances. If all new houses were built with PV roofs and roof-top wind turbines, each new housing development could produce as much electricity as it consumes.

Because total energy needs at present are so much greater than electricity needs, we have to consider in the next chapter how homes and other buildings are to be heated and, in a later chapter, where the energy for transport will be coming from.

Summary and conclusions

Wind power provides a clean and inexhaustible source of electricity. An individual wind turbine near a remote farm, or a cluster near a village, supplying local power, or the occasional wind farm, can add to the interest of the countryside. But there is going to be some limit to the number of turbines that can be installed on land in the UK, if National Parks, areas of scenic or environmental interest, areas of traditional farming, woodland, etc. are to be preserved for wild life, biodiversity and recreation, for escape into the natural world from city life, out of site of slowly-turning turbine blades. The same considerations apply to offshore wind farms. An occasional offshore wind farm of 30 turbines, five miles offshore, can be quite inconspicuous (picture on p.16). But, closer to land, or in greater numbers, such farms will surely begin to disturb sea birds and limit the pleasure of a walk along the wilder stretches of coast, particularly the Scottish Highland coast, where the harnessing of wind power offers large potential. Large offshore wind farms out of sight of land overcome all these objections. One can already begin to see a major new industry developing for the installation, maintenance and, as necessary, replacement of thousands of wind turbines out in the North Sea. In due course, comparable effort may be directed to the development of tidal current and wave power.

The direct generation of electricity from sunlight, using photovoltaic panels and roof tiles, although more expensive than electricity from the wind at the present time, is potentially capable of providing a significant amount of power. Even with existing production methods, PV costs could be substantially reduced by investment in large-scale production.

As wind, wave, tide and sunlight make an increasing contribution to energy supply later in the century, it will become necessary to store the energy from these intermittent sources, during times of surplus, against the times that supply fails to meet need. Small-scale storage, at the level of a village or small town, could be acheived using the system illustrated on p.132. At the national level, renewable energy will have to be integrated into the 'hydrogen economy' (to be discussed in Chapter 3).

THE VALE HOUSE

(Source: *The New Automatous House Design*, Brenda and Robert Vale, Thames and Hudson, 2000)

(Photo copyright: Nick Meers)

CHAPTER 2

THE DESIGN OF BUILDINGS, ENERGY FROM WOOD BIOMASS, HEAT FROM BELOW GROUND

The importance of insulation

About 25% of the total energy used in the UK today goes to heating homes and other buildings [1]. This is energy, and money, totally wasted. It is heating the air around the buildings, due to poor insulation, and this warm air is just blown away with the wind. If the future means depending on renewable energy, it will not be possible to waste heat in this way.

Brenda and Robert Vale's house, at Southwell near Newark in Nottinghamshire, looks quite conventional. From the road you will hardly notice it, set amongst shrubs and small trees, totally in keeping with the old houses of Westgate, the road which runs from the nearby Southwell Minster, with its 12th century nave and 13th century quire. But the Vale's house is insulated to very high standard and a conservatory runs right along the south-west side facing the garden. The house needs hardly any heating, only occasional use of a wood-burning stove at the coldest time of the year.

How is this miracle in energy saving achieved? It is important here to realise that a *perfectly* insulated house would need *no* heating, even if a blizzard were raging outside at −30°C. The body heat of people in the house would gradually raise the temperature inside until they had to open a window, not only for fresh air but also to cool off. The money and energy used to warm conventional houses is simply replacing heat which diffuses out through walls and closed windows, and heat lost in the warm air which escapes, through open or ill-fitting windows and doors.

A second principle involved in the energy-saving Vale house is well known: the warmth which builds up in a conservatory, or greenhouse, or even in a car parked outside, on a sunny day. Glass is transparent to much of the sun's radiation but traps heat [2]. This effect is more marked if a modern form of double glazing is used, in which the inner pane carries a very thin coating, transparent to light but reflective to heat radiation, further reducing energy loss [2]. No real house can be perfectly insulated, so the aim is to make the insulation as good as possible and balance heat loss with heat gained from the sun, coming into a conservatory, or in through large southfacing windows.

The third necessary principle, for house design of this kind, is that the interior of the house must have some mass: a concrete main floor or some concrete or brick internal walls. Then, on sunny days, the mass absorbs heat, preventing the house from getting too warm. Later this heat is given out, keeping the house warm through a few sunless days. Only after a longer spell of grey days in winter is the wood burner needed. The Vale's house has 10 inches (250 mm) of insulation in the walls, 18 inches (450 mm) under the roof and also good underfloor insulation.

Since the material used for roof insulation can be made from recycled newspaper, and good wall insulation requires only a thicker than normal layer of expanded polystyrene between inner and outer walls, this need add only a moderate amount to the cost of building, soon repaid by the saving in fuel bills. Susan Roaf's house (illustrated on p.21) includes all the features we have noted: serious insulation, a conservatory and also internal mass.

The principles of energy efficiency are also being applied in urban building, at Beddington, near Sutton in South London (box on pp.34–35) where, again, there is 300 mm of insulation.

The Vale and Roaf houses were experiments in energy-efficient building and cost was a secondary consideration. However, these principles are also being incorporated successfully into a commercial housing development, at Millennium

The heating arising from sunlight coming through glass is referred to as passive solar heating.

Concrete is not a very environment-friendly material, since a lot of energy is used in its manufacture. Solid walls can alternatively be made from cob (clay and straw), rammed earth, or old tyres filled with earth. Concrete can be made from magnesium carbonate, rather than calcium carbonate, allowing large reduction in carbon dioxide emissions [5].

To be effective as a heat store the mass must be within the insulating envelope which surrounds the house. For the future there is the possibility of using 'phase change' material, rather than sheer mass, to store heat [8].

Modern materials reduce the thickness of insulation needed to achieve the same effect as the 250–300 mm in the Vale and Beddington houses.

Green not far from Southwell. The secret of success depends on attention to detail, such as use of material of low heat conductivity for the ties between inner and outer walls, and closely fitting doors and windows to eliminate draughts. Does this not make the house stuffy? There is a neat solution to this problem, fitted in the Vales' house and at Millennium Green. A small fan draws in fresh air from the outside through a heat exchanger capturing heat from the outgoing warm air. Air is exchanged without great heat loss. The stale air can be ducted from kitchen and bathroom and the fresh air (now warm) ducted to living areas (picture on p.38). The Millennium Green homes also have a set of efficient, vacuum-tube, solar collectors on the roof for water heating [3].

The insulation of buildings will become particularly important if global warming stops or diverts the Gulf Stream, putting Britain into a mini ice age.

Houses built today will still be standing later in the century. It is rather vital to begin to build, from now on, houses with large double-glazed windows, or conservatories, on the southern side, and serious insulation. Houses on an estate can be turned to the sun (plan on p.39) to provide good passive solar heating and a south-facing roof for PV.

PV is very exciting from an architectural point of view as can be seen at the Solar Office in Sunderland (picture p.41). Although most solar panels and tiles are blue at the present time, all colours will be available in the future. The UK Government missed a marvellous opportunity to install PV panels on the roof of Portcullis House, the MP's office building opposite the Houses of Parliament, to show how beautiful these can be, costing no more than the grey aluminium-bronze panels chosen.

Good urban architecture requires a background unity which in the past was imposed by the building materials used, brought to life by individual variation. This is seen in the urban backgrounds of the pictures of Renaissance Italy, in the façades of the burghers' houses along the old canals of Amsterdam and in the Georgian buildings of Bath. Modern architects are freed from structural restraint, and this produces the confused centres of Reading and Basingstoke we see today. PV offers a variety of colour, and allows buildings of great variety of design, but could impose that background unity so essential for a truly satisfying urban environment.

THE BEDDINGTON HOUSING DEVELOPMENT, SUTTON, SOUTH LONDON

The Beddington Zero Energy development (BedZED) is an experiment in urban housing, largely self-sufficient in energy, water supply and sewage treatment, set up by the Peabody Trust in association with the environmental consultants BioRegional and the architect Bill Dunster.

The devlopment consists of seven 3-storey buildings set on an east-west axis, so that their longest sides face south. These sides are glassed-in, to provide a conservatory on each floor. The internal lay-out provides five 4-bedroom town houses, about 20 3-bedroom maisonettes, about 50 one- or two-bedroom flats and about 20 work spaces for offices, or comparable use.

The glass panes of the upper-floor conservatories incorporate a PV area of 780 sq.m. In a separate building a combined heat and power (CHP) plant generates a further 100 kW of electrical power. The CHP plant is fed with wood chips from the maintenance of street trees, parks and gardens in Croydon. These arrive at the Beddington site with a water content of around 50% and are first dried (to a water content around 15%). They are then gasified, that is, heated to high temperature in a restricted flow of air, to produce a mixture of gases (methane, hydrogen, etc.) and leaving an ash of charcoal. The gas flow then has tar removed and the cleaned gas powers an internal combustion engine driving an electric generator.

The value of 'combined heat and power' lies in the fact that the heat generated by the internal combustion engine is not wasted. Heat from the engine cooling system is used to dry the wood chips to 15% moisture content. A heat exchanger captures heat from the engine exhaust flue to supply hot water for the whole development. (The CHP plant is supplied by B9 Energy Biomass Ltd.).

The buildings are insulated with 300 mm (12 inches) of expanded polystyrene (for roof and floors) and 300 mm of Rockwool for the walls (Rockwool is made from molten volcanic rock spun into fibres). A small fan in each residential unit, circulating warm air from around the hot water tank, provides all the supplementary space heating needed.

BedZED feeds power into the grid in summer, withdraws it in winter, to be on balance energy self-sufficient over the year (apart from the input of wood chips) and this includes producing power for up to 40 electric vehicles. The whole development is also largely self-sufficient in water, with rain water collected from the roofs and with the sewage treated in reed-bed tanks to sufficient purity for reuse to flush toilets.

The prominent cowls on the roof provide ventilation in cold weather when close-fitting doors and windows are closed to conserve heat. The incoming cold air is warmed by the outgoing warm air by means of a heat exchanger of the kind illustrated on p. 38.

(Source: Peabody Trust and Bill Dunster Architect, photo: Markus Lyon, BioRegional)

Heating with wood

The central heating in most homes in the UK today is fuelled by natural gas. This is, of course, a fossil fuel so this space heating contributes to carbon dioxide emissions. Also, as we shall see in the next chapter, a steady rise in the price of gas can be expected over coming years. Wood may become increasingly important for space heating. The burning of wood is carbon neutral, since the carbon dioxide released is balanced by carbon dioxide taken from the atmosphere as the trees are growing.

For those living in the country, it is possible to store a stack of logs against the house, and return to a traditional log-burning stove. This is not so easily done in an urban or suburban setting, but there is now an alternative, clean, effortless, way to burn wood illustrated on p.42. Dry waste wood: off-cuts from the manufacture of doors and window frames, or wood crates and pallets no longer needed, can be pulverised and the wood dust compacted to dry pellets. For a block of flats, or for schools and hospitals, central heating boilers are also available burning these wood pellets. For these larger buildings the pellets are delivered to a storage area, from which they can be fed automatically to the boiler. The domestic use of such pellets, illustrated on p.42, is feasible in a home which is really well insulated, and designed in the ways described in the previous section to require little supplementary heating. It would not be practicable for an old draughty house.

A stove of the kind illustrated on p.42 could be fitted with a back boiler for water heating. If we picture a modern house, with passive solar heating, etc. and an efficient solar water heating system on the roof, there is no need for supplementary heating through much of the year. Wood pellets can provide space and water heating backup when this is needed.

We have so far discussed wood pellets from dry waste wood but, of course, wood chips from woodland clearance, from park maintenance, from the branches of felled trees or from short-rotation willow coppice, grown as an energy crop, can all be used as fuel for space and water heating. Wood chips are less suited to domestic use, better suited to larger installations where there are facilities for drying the chips and where the automatic feed can handle the more variable shape and size of the chips (as compared with the pellets).

Combined heat and power

At the older, coal-fired, power stations, the heat produced in electricity generation is wasted. However, as natural gas began to supersede coal in electricity generation it became possible to build smaller generators, situated within the grounds of a hospital or industrial complex, in which the heat produced is usefully distributed for space or water heating. This is 'combined heat and power' (CHP).

Most CHP plants today are gas-fired. However, pellets from dry wood waste, or wood chips from willow coppice and woodland clearance, can also be used as fuel for CHP. This is what is done at the Bedddington site described in the box on pp. 34–35. Wood chips are 'gasified', that is, heated in a restricted flow of air to produce a mixture of gases (hydrogen, methane, carbon monoxide, carbon dioxide, etc.). The gas mixture contains some tar. This is filtered out and the clean gas mixture then fuels an internal combustion engine, to drive a generator producing electricity. About one third of the energy in the wood is converted to electrical power and about two thirds is available as heat. At Beddington the wood-fired CHP plant generates most of the electricity needed for the site and heat is circulated to the flats and town houses for space and water heating.

Southampton is able to draw 15% of the heat needed for its district heating from an aquifer (with water at 76°C) 1.8 km below the city, but heat from this source will only last for 20 years [7].

CHP plants can be conveniently linked to 'district' or 'community' heating in which, as for the wood-fired CHP plant at Beddington, hot water is circulated to flats, town houses, or further, to housing estates, hotels and municipal buildings. Southampton, for example, has over 11 kilometres of thermally insulated piping circulating hot water to the Civic Centre, four hotels, the Royal South Hants Hospital, the Institute for Higher Education and a shopping centre.

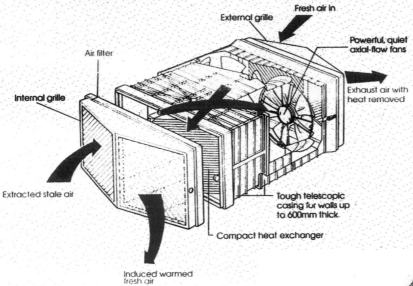

Fresh air in

External grille

Powerful, quiet
axial-flow fans

Air filter

Internal grille

Exhaust air with
heat removed

Extracted stale air

Tough telescopic
casing for walls up
to 600mm thick.

Compact heat exchanger

Induced warmed
fresh air

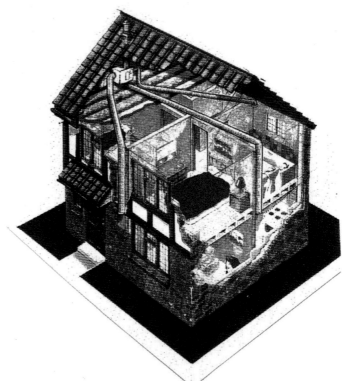

A HEAT EXCHANGER AND THE DUCTING REQUIRED FOR GOOD AIR EXCHANGE

The heat exchanger is seen as the white box in the loft of the house.

(Source: Vent-Axia)

HOUSES ANGLED TO THE SOUTH ON A SMALL ESTATE

(Copyright: Energy Saving Trust)

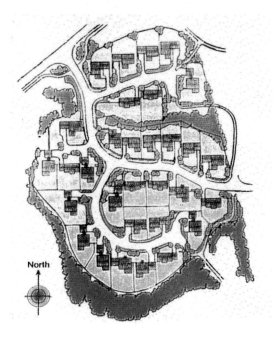

North

We shall see, in the next chapter, that there is another possible use for wood chips in generating transport fuels, and I discuss there the possibility that the waste dry timber available each year could be sufficient for space heating and for generating the extra electricity needed in winter. The pellets (made as described on p.42) from waste dry wood have a low water content and are thus more readily transported and stored than wood chips. They are of uniform size, and thus well suited to reliable, automatic, controlled feed from hopper to burner.

Wind and tidal flow contribute year round, with more wind perhaps in winter. PV and solar water heating fall off in winter, just when more electricity is needed for dark evenings and some heating is needed, even in well insulated buildings.

Wood provides a way to match renewable energy to the seasons. In summer, PV can supply a significant proportion of the electricity needed (and solar heaters can contribute significantly to water heating). As the days became shorter in the autumn, wood-pellet stoves and central heating boilers, and CHP plants could come into action, supplying space heating and the additional electricity and water heating needed during the winter months. The energy stored in trees in summer growth would be harvested for winter use.

At Woking in Surrey, Allan Jones and his team have shown that PV and district heating/cooling CHP can act together to meet a large part of all lighting, heating and air-conditioning needs for a residential and service area (which includes civic offices, two hotels, a conference centre, a leisure complex, a bowling alley, a night club and a multi-storey car park) with the PV largely meeting summer need and CHP playing the main role in winter. This development has shown that CHP district heating, combined with local electricity distribution by private wire (laid down when the district heating pipes were installed) can bring costs down so much that the money saved allows major investment in PV, even at today's price for PV panels. Integrated into a scheme of this kind, PV is not a future dream but is affordable now. At Woking, as at most other CHP plants across the UK today, the fuel is natural gas, but CHP could be wood-fired in the future as it is today at Beddington.

CHP heat is used to cool water by means of a rather clever 'absorption chiller' system.

It is also possible to install CHP within the home, with a WhisperGen unit. About the size of a fridge, this unit can replace a conventional gas boiler, to provide hot water and central heating, and also incorporates a small Stirling engine to generate electricity [4]. Today's WhisperGen units run on gas, but they could easily be adapted to run on waste wood pellets.

Ken Livingstone, the London mayor, has recruited Allan Jones to do for London what he has done for Woking.

In the energy scenario for the UK in 2050, developed in Chapter 5, it is supposed that the energy needed for space heating has been reduced to a third of the energy used to this end today. This will be achievable if the insulation standards for new housing are raised now, to approach the standards of the Vale house and the Beddington development, and if new buildings (homes, schools, hospitals, etc.) are increasingly designed with large south-facing windows for passive solar heating. The insulation of old housing stock can be improved to some extent with cavity wall and loft-floor insulation, by draught-proofing windows and doors, as well as by double glazing.

A well insulated building, as well as being easier to heat in winter, is also easier to keep cool during a hot spell in summer. Usually, in Britain, for a well designed building, sufficient cooling can be achieved with natural ventilation. Where air

SOLAR OFFICE, DOXFORD INTERNATIONAL BUSINESS PARK, SUNDERLAND

(Copyright: Akeler)

conditioning is needed, the power can come from PV, which supplies maximum power at just those times that air-conditioning is needed.

Geothermal energy

There is a further huge store of energy: the heat under the earth's crust. In some parts of the world this heat is near the surface, allowing useful harnessing of geothermal energy. In Britain, one has to go down six kilometres to reach temperatures at which water can be turned into steam, for electricity generation. A research project was started in Cornwall, at the Cambourne School of Mines, which involved drilling two boreholes to a depth of 2 km, opening up fissures in

FILLING THE HOPPER OF A WOOD-PELLET BURNING STOVE

(Source: Renewable Energy Association)

Welsh Biofuels take in each year 10,000 tonnes of waste wood, made up of off-cuts from companies making roof trusses or window frames, as well as old delivery pallets, after they have been reused a number of times. This waste wood is chomped up, pulverised, mixed with steam and pressed into pellets 1/4inch (6 mm) in diameter, 1/2inch (12 mm) long.

the rocks at this depth by forcing water down one borehole at very high pressure, then pumping a steady stream of water down this borehole, which returned warm from the other [6]. Results were not very encouraging. One problem is that drawing energy from the earth in this way is not sustainable. After the heat has been drawn from rocks around the base of the boreholes, it takes thousands of years for these rocks to warm up again, through the slow inward flow of heat from the surrounding earth. Research on this project in Cornwall has now been abandoned, and although this remains a possibility for the future, the harnessing of geothermal energy appears, at present, to offer a solution to energy problems only in more favoured parts of the world.

There is another interesting way of going into the earth for heat, using a heat pump (that is, a motor and compressor, like the unit in a fridge which draws heat from the fridge interior and warms the air behind it). If a borehole is made 120 mm (5 inches) in diameter, to a depth of 50 metres, with a loop of pipe inserted down the hole carrying circulating water, then a heat pump will give out four times as much energy as the energy supplied. In other words, 1 kWh of *electricity* in gives 4 kWh of *heat* out. This is a very efficient way to warm a home, or larger building, using electrical power, as compared with a simple electrical heater (picture on p.44). In summer the heat pump can be run the opposite way to keep the house cool. There is no fear of depleting this heat store, up to 50 metres down, since it is replenished by the sun's radiation falling on the earth's surface.

There is the possibility of making PV panels from cheap organic polymers, such as those being developed by Richard Friend at the University of Cambridge (*New Scientist*, 9 Sept. '06).

At this point, or earlier, a thought may have arisen in the critical reader's mind: are we really going to be dependent only on the sun, wind, tides, etc.? Will scientists not come up with some totally new idea for generating ample energy? One bright hope is nuclear fusion, to be discussed in Chapter 4. But current understanding of thermodynamics (the science of energy) is well developed, both in the physical and biological sciences. We know that matter can be converted to energy, in a nuclear reactor or in a hydrogen bomb, but we also know that generating power in this way carries considerable hazard. The energy 'laws' of physics, established through classical thermodynamics, and more recent study of nuclear reactions, provide a detailed understanding of the evolution of stars, the geological evolution of the earth, and the interactions involved in biological evolution and in living processes. It is most unlikely that any source of energy will be found which does not fall within the framework of these laws. This does not mean that there cannot be brilliant new inventions. For example, chlorophyll (the green pigment of leaves) could perhaps be incorporated into PV design, to mimic the way chloroplasts in green leaves convert sunlight to electricity (which they do, in fact, on a tiny scale). Such design might lead to cheaper and more efficient PV panels.

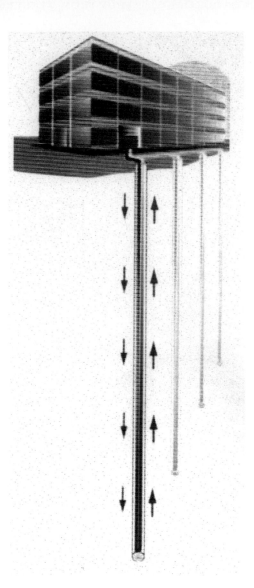

HEAT DELIVERED TO OR DRAWN FROM THE EARTH WITH A HEAT PUMP

This drawing shows the system in the air-conditioning mode with the circulating water going down warm and returning cool. In reverse operation the water goes down cool and returns warm.

(Source: Earth Energy Ltd)

Aside from nuclear power, energy on this planet comes from the sun and the tides. The sun's energy comes as direct radiation, as the force of wind and wave, and stored in the fossil fuels. Energy from the tides is interesting. The tides rise and fall as a result of the gravitational attraction of sun and moon. If we draw energy from tidal power, where does this come from? It comes from the momentum stored in a spinning earth. Energy drawn from the tides slows the earth's spin, making days a little longer. The tides, flowing past rocks and up channels, are consuming energy all the time (generating a small amount of heat). This slowing of the earth's spin is very gradual (the days lengthen by ten seconds every million years) but is perceptible over billions of years. Early in the earth's history, the day was only five hours long. The momentum stored in our spinning earth is so vast, and this slowing so slight, that if we harness tidal power this remains one thing, on this troubled planet, that we do not need to worry about.

Summary and conclusions

Houses of the future will be warm, comfortable and largely self-sufficient in energy, on average over the year. They will draw electricity from the grid in winter but feed back a comparable amount during the summer. Vacuum tube arrays on the roof will provide hot water even on dull days. Wood-burning stoves, and electricity used efficiently to run heat pumps, will provide supplementary space and water heating. Burning, or gasification, of wood at 'combined heat and power' plants, will provide the extra electricity needed in winter and district heating.

Since houses built today will still be standing later in the century, after conventional oil and natural gas supplies have been exhausted, it is important that insulation standards for new buildings be raised now, and that their design incorporate maximum passive solar heating. Then the more limited energy available from renewable sources, such as dry waste wood and chips from woodlands, will be sufficient to meet need.

The emphasis in this chapter on serious insulation, and wood pellets as a fuel for domestic heating and CHP, may seem at present rather extreme. However, this will all become more immediately relevant as the price of natural gas begins to rise. We shall see in the next chapter that such price rise is to be expected.

CHAPTER 3

HYDROGEN

For many years, hydrogen has been seen as the fuel of the future, clean and potentially renewable. This dream is now close to reality, due to the development, mainly by a relatively small company, Ballard at Vancouver, of fuel cells suited to power cars and other road transport. Fuel cells come in many forms. In the type developed by Ballard, a membrane separates air on one side from hydrogen on the other. The membrane is impermeable to the air and the hydrogen, and is permeable only to protons, the tiny nuclei of the hydrogen atoms. Protons carry electric charge, so their flow across the membrane generates electricity (box on p.50).

In the internal combustion engine, whether petrol or diesel, the fuel burns at high temperature in the cylinders and the exhaust from a car tailpipe includes the products of combustion: carbon dioxide, nitrous oxides and, in the case of diesel, minute particles which are carcinogenic (cancer causing). The proton-transfer fuel cell operates at low temperature and the exhaust is pure water. Fuel cells can be more efficient than internal combustion engines, needing lower energy input to produce a given power output.

Ballard and other companies are turning the fuel cell from a hopeful idea to a reliable power source, at present still relatively expensive. A hydrogen-powered fuel cell scooter is being developed by Aprilia in Italy. The first hydrogen-powered cars are expected in the showrooms by 2010. The major car manufacturing companies are competing to be first in the field. The fuel cell powers an electric motor which drives the wheels.

There are problems, one being the storage of hydrogen. Other fuel gases, like propane and butane, can be liquified at moderate pressure. Propane is widely distributed today in the heavy cylinders used by caravaners. Light cans of butane are used by campers for light and cooking. Hydrogen, the lightest of all the gases, can only be liquified at very low temperature. The hydrogen-powered scooter will store hydrogen as a compressed gas in a light-weight carbon-fibre tank, carrying enough hydrogen for travel at 55 km per hour for 190 km. A car is going to need to store much more hydrogen than this to allow it to travel a few hundred kilometres between refills. It is possible that, in the future, hydrogen might be stored at high density in a tank filled with minute tubular fibres of carbon. This is a technology as yet unproven, but meanwhile Amory Lovins has shown that today's cars, though excellently designed in many ways, are poorly designed for energy efficiency [1]. A much lighter car, made from carbon-reinforced plastic rather than steel, strong in a crash, aerodynamically designed for minimal air resistance, would require only half the energy used in today's cars to travel the same distance at the same speed. A car made to this design may be able to store sufficient hydrogen, as a high-pressure gas, for adequate milage between refills.

For this scenario, it will also be necessary to set up a network of hydrogen filling stations. It is difficult to see the hydrogen-powered scooter becoming a commercial success before a network of hydrogen filling stations is established.

Buses have been the first vehicles running on hydrogen. They are large enough to accommodate hydrogen-storage cylinders, return to one depot for refuelling, and their use contributes significantly to cleaner city air. Hydrogen-powered buses have been running, in an experimental way, for some time now, in Vancouver and other cities, and more recently in London.

There are problems of safety in the use of hydrogen. It must be handled with care, but so must petrol, yet it has been possible to develop a safe petrol distribution system.

The production of hydrogen

There is now a further vital question: where will the hydrogen come from? Hydrogen can be generated simply in two ways: by passing an electric current through water (electrolysis, see p.50) or by interaction of steam with natural gas to produce hydrogen and carbon dioxide (a process known as reforming described in the box on p.54).

The first commercial fuel cell system to be installed in the UK is at Woking, in Surrey. Supplied by UTC Fuel Cells, Connecticut, USA, it provides electricity and heating for the recreation centre. This is a stationary system (a unit five metres long, three metres wide and three metres high, much too large to fit in a car). It includes a stack of fuel cells and also, most interestingly, a unit which carries out the chemical processes described in the box on p.54, converting natural gas to hydrogen.

We can perhaps now begin to see a pattern for the future: as hydrogen–powered cars come into the show rooms, and as they become more affordable, filling stations across the UK could install hydrogen–reforming units of the kind now operational at Woking, fueled by natural gas.

The 'hydrogen economy' is still, to some extent, a fantasy. In this chapter we are exploring ways in which this fantasy might become reality. For a cautionary note see p.62.

As more cars, buses and vans, begin to run on hydrogen, the air in cities will become cleaner until one day, the only cars, delivery vans or other vehicles allowed into towns will be those which emit only water vapour. So far, so good. But, although this reduces dependence on oil and eliminates all the pollution associated with the internal combustion engine, the whole economy still depends on fossil fuel (natural gas) and at each filling station, where natural gas is being converted to hydrogen, carbon dioxide is being released into the atmosphere. The problem of climate change has not been solved.

ELECTROLYSIS OF WATER

When an electric current is passed through water (H_2O) some of the water molecules are broken down to produce hydrogen and oxygen as gases (a process known as electrolysis). Energy is needed to achieve this separation (the electrical energy of the current). This energy is stored in the gases. A spark is enough, in a mixture of hydrogen and oxygen gases, to release this energy in an explosion. The two gases come together again to produce water.

The proton-exchange membrane fuel cell

In a proton-exchange fuel cell, the energy stored in hydrogen and oxygen is released in a controlled way. The cell consists essentially of a membrane separating hydrogen on one side from oxygen on the other. The membrane is permeable only to the tiny nuclei of the hydrogen atom: the protons.

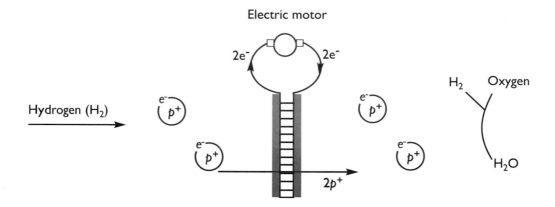

The hydrogen atom is made up of a tiny central proton (shown in this diagram as p^+) with a circling electron (shown in this diagram as e^-). The trick of fuel cell design is to make the membrane permeable to the protons but not to the larger hydrogen atoms.

Oxygen has an affinity for protons and draws them through the membrane. This generates an electrical voltage across the membrane which, in turn, draws electrons along the only route available to them: up one wire, through the motor and back through the other wire. The flow of electrons through the motor is an electric current, so this turns the motor. The blue lines represent layers of platinum, acting as a catalyst.

A HYDROGEN-POWERED CAR

(Source: DaimlerChrysler UK)

The car shown here, an experimental model developed by DaimlerChrysler, has a range of 600 km between fuel refils. But this is achieved by carrying the fuel as methanol (methyl alcohol) and building into the car a chemical plant converting methanol to hydrogen, a rather complicated solution to the fuel storage problem. An experimental car developed by Ford carries hydrogen as a compressed gas, but has a range of only 200 km. A light-weight and more aerodynamic car of comparable size built, as Amory Lovins puts it, 'more like an aeroplane than a tank', yet still ultra-strong in a crash, could achieve a more respectable range of 400 km between refills. The Honda FCX, in fact, is expected to have a range of 350 km and a top speed of 150 km/hour, running on compressed hydrogen.

National and global reserves of oil and gas

UK oil production from the North Sea peaked in 1999 and gas production in 2000. Since then the North Sea fields have been in decline (Department of Trade and Industry statistics, 2005). North Sea oil and gas will last till about 2020, with increased dependence on imported gas from 2005, and on imported oil from about 2010.

The global situation is more difficult to assess. Oil which is easily accessible, such as the that under the North Sea or under the sands of Arabia, which can be simply pumped to the surface, or sometimes wells up under subterranean pressure, is now called 'conventional oil'. Estimates of the total remaining range from a trillion to over two trillion barrels (see *The End of Oil*, Paul Roberts, Bloomsbury, 2004). Nearly one trillion barrels have been used already, and if little more than a trillion remain we are coming close to 'peak oil', that is the year at which extraction reaches 50% and global production peaks. If over two trillion barrels remain, global production may not peak till the 2030s.

The graphic on p.55 is based on a mid-position, favoured today by most analysts and assumes that around 1.5 trillion barrels remain. If this is the true position, then the peak of global production will come some time around 2020. The curves on p.57 show, in more detail, projections in which production peaks as early as 2010 from the Association for the Study of Peak Oil (ASPO).

Also shown on p.55 are estimates for 'conventional' natural gas reserves. With an estimated 21% extracted so far, the peak of global gas production lies further ahead than peak oil, but gas reserves (equivalent in energy to 2 trillion barrels of oil) are little larger than current reserves of conventional oil. Large reserves remain of 'unconventional' oil, that is, oil locked in tar sands and shales, from which extraction is more difficult and costly. There are also large reserves of natural gas, locked in sea floor sediments, again difficult to extract. Finally, there are vast remaining reserves of coal.

The possibility of continuing to use these large reserves of unconventional oil and coal, and trapping, or sequestering, the carbon dioxide so produced, preventing its escape into the atmosphere, will be discussed in more detail in Chapter 4, but we can introduce this topic here. It has been suggested that reforming plants could be sited close to natural gas fields and the carbon dioxide produced in the reforming process returned below ground to the emptying field [2]. It would be possible, in principle, to install reforming plants at the points where North Sea gas reaches shore and, later, at the points where imported natural gas enters the UK, and pump the carbon dioxide to the empty oil and gas fields under the North Sea. If storage of carbon dioxide in old oil and gas fields can be made safe for the long term, then natural gas could still be used for hydrogen production *without* this contributing to global warming.

At this point we have come to an some interesting conclusions: by the time conventional oil reserves run low, around mid-century, the need for oil will be much reduced, since land transport, at least, will be mainly based on hydrogen. The supply of hydrogen is assured, as long as reserves of natural gas last. If global warming is seen as an increasingly serious problem, investment could be made in reforming plants, with return of carbon dioxide below ground. The only problem with this scenario (as we can see from the box on p.55) is that natural gas reserves are only a little larger than conventional oil reserves. As global oil production peaks and begins to decline, there will be increased global consumption of natural gas, used directly or converted to hydrogen. Natural gas reserves may last little longer than oil reserves. We have to consider now other ways in which hydrogen can be generated to maintain future energy supply.

A group of energy companies are in fact planning to install a large reforming plant at Peterhead, in Scotland, producing hydrogen from North Sea natural gas, with return of carbon dioxide to a nearby oil field to enhance oil extraction (see p.76).

THE PRODUCTION OF HYDROGEN FROM NATURAL GAS

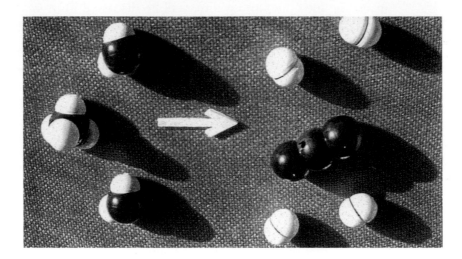

The natural gas is mixed with superheated steam and passed through tubes packed with a nickel catalyst at 760–810°C. The main component of natural gas is methane (CH_4) and the output from this 'reforming' process is a mixture of H_2, CO and CO_2:

$$CH_4 + 2H_2O = CO_2 + 4H_2$$
$$CH_4 + H_2O = CO + 3H_2$$

The carbon monoxide (CO) must now be converted to carbon dioxide (CO_2) by passage over a chromium and iron oxide catalyst at 370°C and then copper oxide catalyst at 230°C

$$CO + H_2O = CO_2 + H_2$$

When this gas stream is cooled the steam condenses out to leave a mixture of hydrogen and carbon dioxide. A catalyst is a metal which facilitates a chemical reaction (while itself remaining unchanged).

Incidentally, the picture in this box shows the two molecules mainly responsible for global warming, carbon dioxide on the right, methane (CH_4) on the left.

(Source: BP, Grangemouth)

FOSSIL FUELS: REMAINING RESOURCES

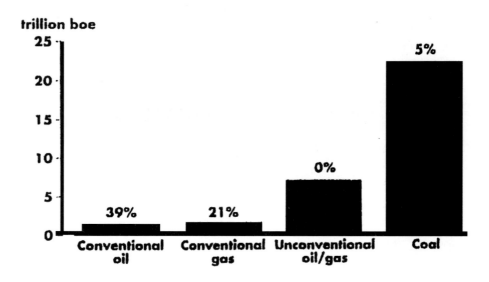

% = total resources already consumed

The estimated reserves of conventional natural gas (energy equivalent to 2 trillion barrels of oil) are little larger than mid-range estimates for reserves of conventional oil (1.5 trillion barrels). Very large reserves remain of unconventional oil in tar sands and shales, and unconventional gas in sea-floor sediments. Vast deposits of coal remain.

boe = barrels of oil, or the energy equivalent for other fuels.
1 barrel of oil = 0.13 tonnes = 160 litres

(Source: Adapted from a graphic in Managing transition, energy supplies in the 21st century, Mark Moody-Stuart, paper given at Sanderstolen Conference, Feb. '98).

The production of hydrogen from renewable sources

We noted earlier in this chapter (box on p.50) that hydrogen can be made from water by electrolysis, a process which does not create energy, but which converts electrical energy to energy stored in the hydrogen. Iceland is in the happy position of having a small population, abundant hydro-electric power from waterfalls and also geothermal energy from which to generate electricity. Iceland is aiming to be the first country to power cars, buses and even fishing boats with hydrogen generated by electrolysis The position is not so favourable for the UK.

We have estimated earlier that about 20% of the total energy consumed today in the UK could be generated from wind, tidal currents, wave power and PV in and around Britain (see Chapter 1). If buildings were well insulated, with maximum use of passive solar heating, then pellets from waste wood, and wood chips, could meet the need for space heating, which today accounts for 25% of total energy consumption (see Chapter 2). The renewable energy sources and efficiency measures we have discussed so far still leave a shortfall of around 55% on present energy use. This is the energy from oil and natural gas used today for transport and for industry, including the production of plastics, fertilizer and other chemicals. The rest of this chapter, and Chapters 4 and 5, are concerned with the question of how best to meet this shortfall in the long term, after conventional oil and natural gas reserves are exhausted.

In principle, the UK offshore wind programme could be expanded to supply electricity for conversion to hydrogen, in addition to electricity for direct use. To meet this 55% shortfall by generating hydrogen from wind power, a total of at least 120,000 3MW wind turbines would be needed [4]. The large (3MW) turbines have to be set about 500m apart to make best use of the wind, that is, about 4 per square kilometre. 120,000 turbines would occupy 30,000 sq.km., a strip 10 kilometres wide running for 3000 kilometres. It seems quite unrealistic to imagine the offshore wind programme being extended in this way (over the present century, at least). This would have to be additional to the installation of five large turbines per week required for electricity supply (as discussed in Chapter 1).

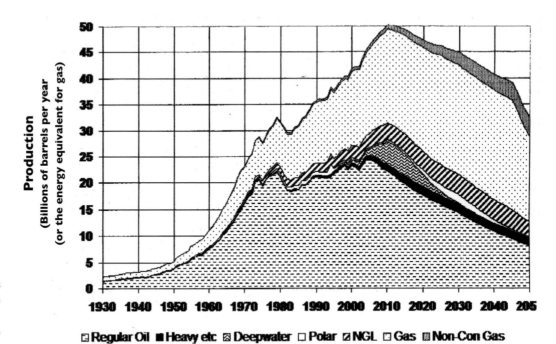

ASPO projections
(NGL = natural liquid gas)

(Source: www.asponews.org)

It is possible that hydrogen could be made directly from sunlight by panels of titanium oxide or nano-crystalline metal oxides (see *Green Futures*, Autumn, '03 and www.hydrogensolar.com). Possibly the algae *Chlamydomonas reinhardtii* could be engineered to do this (see Paul Roberts, *The End of Oil*, Bloomsbury, 2004).

Alternatively, it is possible that PV panels could be set up in desert areas with abundant sunshine, in North Africa for example, and the electricity so generated brought to a source of water. There hydrogen could be generated and piped to the industrial world, or possibly transported in liquid form in super-insulated tankers. Where the electricity is brought to the sea, there will be the further need for desalination plants to produce the deionized water needed for electrolysis [3]. Calculation shows it would need an area of desert of at least 3000 sq.km. covered in PV panels to supply the 55% shortfall in UK energy needs [5]. This can be seen as a strip 1 km. wide running for 3000 km. or as a strip 10 metres wide running for 300,000 km. However we look at it, a huge number of turbines and/or a large area of PV would be needed to meet the 55% shortfall.

In the next chapter we shall be considering nuclear power and also the possibility of using coal and unconventional oil in an environmentally benign way. In a later section of this chapter we shall see that hydrogen can also be generated from wood chips. But before going on to discuss these alternative ways of meeting the shortfall, let us look at the problem from a different perspective and ask whether this much energy is really *needed*, whether, with thoughtful planning for the future, energy demand could not be reduced significantly.

As briefly mentioned at the start of Chapter 1, it may not be possible, using renewable energy, of necessity gathered over large areas, to sustain the wasteful use of energy seen today in the industrial countries. Ernst von Wiezsacker, Amory and Hunter Lovins, in their book *Factor Four* [1] suggest that, with good industrial design, life could be just as comfortable in the industrial world using only one quarter of the resources used at present. Very large savings in energy are possible without any sacrifice in the quality of life. For example, at Kalundborg, in Denmark, factories are grouped together in such a way that the *waste* from one factory is exactly what is needed as a *resource* for the factory next door. This grouping saves a lot of energy otherwise needed for transport and waste disposal. To take another example, European legislation to be introduced during the next decade will require manufacturers of cars, domestic appliances, photocopiers, computers and other consumer goods, to take back these products at the end of their useful life and dispose of them in an environmentally acceptable way. This will lead to the products being designed, from the start, for ready disassembly into their component parts for later recycling. This will much reduce the energy needed to make a new product.

In addition to such immediate ways of reducing industrial energy use, there is the possibility of quite revolutionary developments in technology, particularly through study and imitation of the way the natural world achieves miracles of engineering. Thus fibres might be made in the future with the very high strength-to-weight ratio of the threads of a spider's web using little energy. Genetically-engineered bacteria could be used to carry out chemical reactions at ambient temperature, which today require high temperatures and metal catalysts.

see: www.biomimicry.org

Reducing industrial energy use by a factor of four may be an over-optimistic aim, but reducing this component of total energy use by a factor of two could be achievable.

As cars, buses and trucks, are increasingly fuelled by hydrogen, this will reduce the energy needed for transport, since fuel cells are roughly twice as efficient as internal combustion engines. We shall consider, in Chapter 5, ways in which the level of road transport could be reduced. In the scenario developed in that chapter, for the UK in 2050, both industrial and transport energy use are supposed to have fallen to half current values. This is a future very different, perhaps, from the one that most people expect.

Hydrogen from woodlands

There is now another possibility to consider. The idea of forests of trees, rather than of forests of wind turbines, supplying energy for the future, may at first sound a little far-fetched, but in a report produced for the UK Department of Transport in '03, Nick Eyre and co-authors are enthusiastic about using woody biomass to produce hydrogen [6]. We have seen (p.37) that wood chips can be 'gasified' to produce a mixture of gases, including hydrogen and methane. Methane can be converted to hydrogen (p.54). The gas mixture from gasified wood chips also contains carbon dioxide, but this can be separated out and released into the atmosphere. This is carbon dioxide that was taken up from the atmosphere during the growth of the trees. Nick Eyre and co-authors calculate that if 4 million hectares were set to willow coppice producing 20 oven dried tonnes/hectare/year, this area would be sufficient to fuel road transport in the future at the current level of road use [6].

The UK needs more woodland in the upland valleys to act as a sponge to reduce flooding.

But now we have to ask whether such a large area of coppice would make for an attractive landscape? Why not consider a bolder vision of reforesting Scotland, Yorkshire, Cumbria and other suitable areas of Britain, with mixed woodlands of native broadleaf trees and conifers? This would provide oak, ash, birch, beech and pine for building construction and manufacture of furniture and wood chips from forest clearing, from the branches of felled trees and waste in sawn timber production. The multi-functional aspect of woodland use could make such development as economically viable as willow coppice.

The area of forest and woodland in the UK today is 2.8 million hectares (Forestry Facts and Figures, Forestry Commission, 2002). The annual yield from a roughly equal mix of broadleaf and conifer may be estimated to be around 6 oven-dried tonnes/hectare [7]. About half this yield is available as sawn timber, half could be available as wood chips for hydrogen production. If the UK woodland and forest area were increased just under threefold, to 8 million hectares, the hydrogen produced in this way could make a major contribution to road transport needs, particularly if the level of road transport could be reduced [7]. This increase in forest area would bring all the benefits of timber and chip production and also provide areas of woodland for recreation. A future Britain with three times the present woodland area and a reduced level of road transport seems a rather attractive solution for the future. An increase in forest area to 8 million hectares would bring the woodland area to 33% of the total area of Britain, comparable with the current figure for Italy (34%) and only a little above the current figures for Germany and France of 31% and 28% (Forestry Facts and Figures, Forestry Commission, 2002).

Timber-frame houses (already widespread in Scandinavia, Canada and the US) are becoming increasingly popular in the UK, because of their properties of good insulation and rapid construction. Why not grow more timber in Britain close to where it is needed?

In the scenario for the UK in 2050, developed in Chapter 5, it is supposed that the woodland area has been increased to 8 million hectares as proposed here, significantly reducing timber imports. Waste timber from houses knocked down in urban renewal, from disused pallets and packing cases, etc. is regarded as a valuable resource. Pellets made from this waste timber make an important contribution to space heating needs and generation of the extra electricity needed in winter [8]. In this scenario, wood chips from the clearance of forest, parks and gardens and from the waste in felling and sawing timber are all gasified to hydrogen, making a major contribution to road transport needs.

Aviation fuel

Highly-qualified engineers from 11 European countries are coming together to design a hydrogen-powered plane in the CRYOPLANE project. The pessimistic view of the future of hydrogen as an aviation fuel presented here will have to be revised if this project proves a success.

Although it is reasonable to expect that, by mid-century, the hydrogen economy will be well established for land transport, and perhaps for marine transport, with hydrogen replacing the oil and petrol used today, the future for aviation fuels remains uncertain. Liquid hydrogen is a great fuel for rockets, but whether it can be safely, and economically, used for commercial flight remains to be seen. To keep hydrogen at very low temperature, in liquid form, it must be stored in a double-walled tank with a vacuum between the two walls (like a picnic vacuum flask, but larger and made of metal, not glass). Designs for planes fuelled by liquid nitrogen place the fuel tank in the fuselage. It may be that ethanol, or other liquid fuels derived from plant material, may prove more suitable than hydrogen for aviation.

Liquid aviation fuel could be made from coal, but if coal were used in this way this would still be adding to carbon dioxide emissions.

The UK could import biofuels: oil from rape or other seed crops, ethanol or other liquid fuels, derived from plant material grown in developing countries, but this may not be an ideal solution. If the global energy situation reaches a point where large corporations, such as Shell or BP, invest in wind farms or PV panels in barren, sparsely-populated areas of the world, there is ample desert available without encroaching on land suitable for crops. Growing seed oil or other biofuel crops, on the other hand, requires good agricultural land and is thus in competition with the growth of food and fibre crops (cotton, flax, hemp, etc.). It is going to prove difficult to grow enough food for a global population expected to increase to some nine billion by 2050 and this will put serious restriction on the growth of biofuel crops.

The import of biofuels into Europe could lead to further loss of rainforest to palm-oil plantations (*New Scientist*, 19 Nov '05).

In their book, *Global Warming and Social Innovation* (Earthscan, 2002) Marcel Kok and co-authors consider two future scenarios for the Netherlands, one based on the dynamics of the free market, with low priority given to concern about nature and the environment, the second very similar, in its lifestyle, to the scenario I am developing in this book for the UK. Both their scenarios rely heavily on biomass, or biofuels, imported from Eastern Europe, Latin America or sub-Saharan Africa, a solution to the energy problem that seems questionable, from an ethical

A SMALL PLANE POWERED BY ETHANOL

(Source: Max Shauck, Baylor University, Waco, Texas)

Ethanol (common drinking alcohol) is made by fermenting grain or other sugar, or starch, crops. In the US 6% of the maize crop is used for ethanol production. This small plane is powered by a piston engine, but plant-derived fuels can perhaps be developed suited to jet engines. The jet fuels used today remain liquid down to −40°C, a temperature at which vegetable oils and biodiesel oil are solid. Possibly, in jets of the future, the fuel tanks and supply lines could be insulated and heated to overcome this problem, or additives could keep the fuel liquid at these low temperatures.

viewpoint. If, indeed, countries in these regions can sustainably, and without further environmental destruction, meet all their own needs for food and energy and then produce surplus biofuels for export, well and good. Otherwise, this could prove yet another example of the developed world exploiting the resources of developing countries (see Chapter 6).

The picture of a small plane fuelled by ethanol (above) raises an interesting point: some analysts believe that hydrogen will never prove a suitable fuel even for land transport, that hydrogen-powered cars will remain expensive, that limited global reserves of platinum will curtail production of fuel cells, that no substitute for platinum will be found as a catalyst, and that the difficulty of onboard storage of hydrogen will limit range between refilling. This seems a rather pessimistic view, but it is quite possible that land transport in the future could be based on liquid biofuels and the internal combustion engine, rather than hydrogen and fuel cells. This would be a less satisfactory solution, in that there would still be nitrous oxide emissions from the internal combustion engines, but a solution which would be totally acceptable from the global warming viewpoint. Methods may be

Current research at Oxford University suggests biological enzymes could perhaps replace the costly platinum catalyst in fuel cells (www.isis-innovation.com).

For production of ethanol from wood see www.iogen.com.

developed for ethanol production at high yield from woody biomass and the efficiency of conventional engines could be improved to match the efficiency of hydrogen-powered cars. The coppice or woodland areas I have allocated to hydrogen production would then be devoted instead to ethanol production. This does not alter the fact that the area of the UK available for hydrogen or biofuel production is limited, and that the import of biofuels may prove ethically questionable.

Even with the greatest feasible improvement in the efficiency of energy use, it seems that it will prove difficult to meet future energy needs from renewable sources. We consider, in the next chapter the possibility of continued use of nuclear power, or of coal, with carbon capture and storage. In Chapter 5 we discuss ways in which the level of road transport could be reduced, with improvement in the quality of life.

Summary and conclusions

The most probable scenario is that, over the next 20–30 years hydrogen-powered cars will become affordable and buses and lorries will be increasingly powered by hydrogen. The air will be clean and by the time the oil runs out around 2050 we shall no longer need petrol and diesel fuel.

The hydrogen will come initially from natural gas. The reforming plants for this conversion could be sited close to the gas supply fields, with carbon dioxide returned to the emptying gas chambers if the problem of global warming becomes acute.

This looks good but, even if this transformation is achieved, dependence on oil will have been replaced by dependence on natural gas. Reserves of natural gas (although some remains perhaps to be discovered) are basically no larger than reserves of oil. This is no long-term solution. To supply hydrogen, in the long term, for industrial and transport needs in the UK alone, would appear to require vast areas of wind farms and/or PV panels in deserts, or sparsely populated areas of the world, with an extensive new infrastructure of power lines, electrolysis plants and hydrogen pipe lines. An attractive alternative would be to set large areas of the UK to willow coppice or woodland.

Liquid fuels from plant material may prove more suitable than hydrogen for aviation, but where fuels are derived from plants their production is in competition with food production, in a world seeking to feed a population rising to 9 billion.

CHAPTER 4

NUCLEAR POWER AND COAL. ENERGY FROM WASTE

Nuclear Power

Nuclear power is unpopular at the present time. The accident at Chernobyl in 1986 showed how disastrously things can go wrong, leaving large areas of the Ukraine and Belarus highly radioactive. Fifteen years later the sheep on around 400 farms in the North of England were still subject to restriction orders because their meat contained significant levels of radioactive caesium. The leaking, entombed ruin of the power station remains a serious hazard. However, as the effects of global warming become more apparent and as oil and gas begin to run out, the nuclear option is again being given serious consideration.

Within the core of a nuclear 'reactor', uranium in the fuel elements generates heat, which is used to produce steam which drives turbines, which generate electricity. The core is contained within a casing, which may be steel, up to 225 mm (9 inches) thick, or a thinner steel shell, surrounded by concrete which may be up to 3 metres thick. During the working life of the reactor the fuel elements are replaced as necessary to maintain the reaction and ongoing heat supply.

The heat from the core must be brought out, by circulating gas or fluid, to heat-exchangers where the steam is generated. Ducts, or pipes, must pass through the concrete and steel shells into the core. Within the core, radiation causes steel to become brittle with time, welds can become unsafe and cracks develop. During the operation of the reactor, radiation levels build up within the core and in the casing, and these levels remain high even if the fuel elements are removed. Maintenance becomes increasingly expensive and hazardous until after 50 years

DECOMMISSIONING THE NUCLEAR POWER STATION AT BERKELEY, NEAR BRISTOL

The two Magnox nuclear reactors at Berkeley began operation in 1962, producing together a rated 276 MW of electrical power [1]. They were shut down in 1988/9 after 27 years of operation. Berkeley was the first commercial nuclear power station in the UK to be decommissioned, the work being carried out by British Nuclear Fuels Ltd (BNFL).

The first stage of decommissioning was the removal of the fuel elements, carried out over a period of three years. Removed in batches, these fuel elements had to be stored on site for 100 days in 'cooling ponds', that is, in large water-filled concrete tanks the combined size of an Olympic swimming pool (these tanks were installed when the station was first built to cool batches of spent fuel during the working life of the reactor). The spent fuel elements were then returned to Sellafield for storage or reprocessing. 'High-level' waste, that is, the highly-radioactive fuel elements, had all been removed from the site by 1992.

After this, the cooling ponds themselves were removed. This involved scraping 50 mm of contaminated concrete from the inner surface of the tanks. This was low-level (less radioactve) waste and was sent to the storage site for low-level waste at Drigg in Cumbria. Radioactive caesium leaches out from the elements during the cooling period. The water from the tanks had to be filtered through a resin, which absorbs the caesium, and the contaminated resin is being stored on site as intermediate-level waste. This waste is embedded in concrete for storage, within a stainless steel casing, and will be stored on site until an intermediate-level waste-storage site is established in the UK. The outer, uncontaminated, parts of the cooling tanks were then demolished and this part of the site was infilled and landscaped. The building in which the turbines had been housed was stripped, prior to its demolition, and the turbines and other equipment reconditioned and refurbished for future use elsewhere. This policy of recycling materials, after extensive monitoring to check for radioactive contamination, was followed in all areas. The heat exchangers remain on site, presenting no hazard (they are seen as rusty, cigar-shaped steel containers in the picture in this box). The highly radioactive cores of the two reactors, with their steel and concrete casings, remain within the two large buildings seen in this picture. One of the surrounding low buildings stores the intermediate-level waste. This store is built with shielding within the walls, reducing radiation outside the building to very low levels.

All this will remain, with no foreseeable hazard, for 100 years, with no maintenance required beyond occasional inspection. After 100 years, the radioactive cobalt in the steel casing of the reactors will have decayed to low levels and it will be relatively easy to move in with remote-controlled equipment and totally clear the site for alternative use. The whole decommissioning process is being carried out very carefully by BNFL, though of course at some cost.
By the year 2100, the Berkeley site may be fully decommissioned, but we have to remember that some of the radioactive material has not really been disposed of, but only transferred to Sellafield. In addition to classifying waste into high-,

intermediate- and low-level components, based on the level of emitted radiation, we must now distinguish between long-life and short-life radioactive material. The radioactive cobalt in the irradiated steel casing is of relatively short life, decaying to half its present radiation level after about 5 years, to 1/4 after 10 years, to 1/8 after 15 years and hence to a very low level after 100 years. The caesium in the intermediate-level store, on the other hand, has a half-life closer to 30 years and will only have decayed to 1/4 of the present level after 60 years and to 1/8 after 90 years. Many of the by-products of nuclear reactions have half-lives of centuries or longer.

These long-life components are the Achilles heel of the nuclear industry. What is to be done with them: should they be stored for centuries above ground at Sellafield, stored in a deep cave in the stable rocks of Greenland, divided into small portions and sent on rockets into outer space, or placed at the target area of a cyclotron to break them down into short-life components? Until this problem is solved it is difficult to estimate the true cost and the true risk of ongoing reliance on nuclear power.

(Source: British Nuclear Fuels Ltd)

(or a shorter time in the case of early reactors) further operation of the reactor is uneconomic. It must be closed down and decommissioned (box on pp.66–67). Only partial decommissioning is possible initially. The reactor must then be left for 100 years or so, for radioactivity levels to fall, before the site can be fully cleared.

The essentially unknown cost of the final decommissioning, 150 years after the reactor is built, makes it difficult to estimate the true cost of nuclear power (see audit on p.141). No one knows yet how to ultimately dispose of long-life radioactive waste or whether this can be safely stored for some 100,000 years.

There is the possibility that nuclear power may, one day, be generated in a *fusion* reactor. This would be a type of reactor in which hydrogen molecules fuse to form helium, with release of energy. This is the reaction which powers the sun. Much research has already gone into attempts to control and harness this fusion reaction, with only moderate success. We cannot confidently rely on energy from fusion becoming available, even in the distant future. The core of a fusion reactor has to be held at 100 million °C (many times greater than the temperature at the centre of the sun) contained within an electromagnetic field.

Our present day nuclear reactors are *fission* reactors, that is, the energy is derived from the spitting of uranium or plutonium atoms into smaller atomic fragments. Today's reactors use enriched uranium as the fuel (that is, natural uranium with its isotopic composition modified to make it more fissionable). In the reactor, plutonium is formed from uranium, and the spent fuel can be reprocessed to separate the plutonium. This plutonium can then be used to enrich the next batch of fuel (now called MOX, mixed oxide fuel). In principle, by this means and by using 'breeder' reactors designed for plutonium production, the nuclear industry could continue to generate its own fuel. The problem is that this is all very expensive and dangerous to do. Transporting spent fuel from Japan for reprocessing at Sellafield, then later shipping the MOX to Japan carries the risk of accident on rail, or at sea, and the possible risk of its capture by terrorists competent enough to make a primitive nuclear bomb. The reprocessing at

The safety of waste stored at Drigg could be threatened by sea level rise (*Independent on Sunday*, 19 June '05).

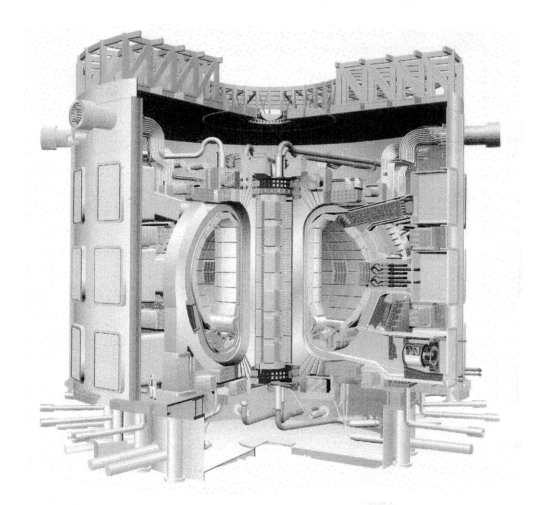

ENERGY FROM NUCLEAR FUSION

There are plans to form an international consortium to build the experimental fusion reactor illustrated here, by 2012, at an estimated cost of £2.45 billion. It would be interesting to see £2.45 billion invested also in the development of cheap PV roof tiles. Then, in 2012, it would be possible to judge which approach looked most promising for the future.

(Source: US Department of Energy)

Sellafield releases radioactive material into the sea and into the atmosphere (see p.71) and the site is also extremely vulnerable to terrorist attack, with release of radioactive materials possibly more disastrous than that from Chernobyl.

There remains also the ever-present danger of further operational accident on the scale of Chernobyl and the problem of how to eventually dispose of the increasing amount of this highly dangerous substance plutonium which is constantly generated in fission reactors.

All these considerations apply, with greater force, where nuclear power stations are built in politically unstable parts of the world, or sited in regions prone to earthquakes.

Nuclear power, in the UK today, contributes one quarter of electricity needs (see pie chart on p.19) that is 5% of total energy needs. Building new nuclear power stations, to replace those being decommissioned over the next 20 to 30 years, would retain all the hazards involved while still contributing little to total energy needs. To rely more completely on nuclear power would require an increase in construction beyond replacement of the present stations.

It is possible that the UK Government, foreseeing the exhaustion of North Sea oil and gas, realising the limited extent to which energy from renewable sources in and around the UK can replace the total fossil fuel used today and seeing how dependent Britain will become on imported natural gas, will feel it prudent to build new nuclear power stations, for national security in energy supply.

Enthusiasts for nuclear power point out that the design of the Chernobyl reactor was one which would never have been accepted as safe in the West and claim that the next generation of reactors can be designed to be very safe, both in normal operation and against terrorist attack. These new reactors will produce much less waste and, if the uranium is enriched by centrifugation rather than by the earlier diffusion method, it is estimated that they will generate the energy used in their construction, maintenance and eventual decommissioning within a

**RADIOACTIVE
EMISSIONS
FROM SELLAFIELD**

(Source: *The Ecologist*, Sept, '01)

Into the atmosphere go wastes such as:
tritium, carbon-14, sulphur-35, argon-41,
krypton-85, cobalt-60, strontium-90,
ruthenium-106, antimony-125, iodine-129
and 131, caesium-134 and 137, plutonium
239, 240 and 241, americium-241 and
curium-242.

Into the sea and river goes a cocktail even
longer in the listing: tritium, carbon-14,
sulphur-35, manganese-54, iron-55, cobalt-60,
nickel-63, zinc-65, strontium-89 and 90,
zirconium-95, niobium-95, technetium-99,
ruthenium-i03 and 106, silver-110m,
antimony-125, iodine-129, caesium-134 and
137, cerium-144, promethium-147,
europium-152, 154 and 155, neptunium-
237, plutonium-239, 240 and 241,
americium-241, curium-242, 243 and 244,
and uranium.

few years of operation [3]. Nuclear power can thus produce a great deal of net energy without contributing to carbon dioxide emissions. A nuclear reactor operating at high temperature could provide the energy for hydrogen production from water as an alternative to production by electrolysis.

The problem with these sites is that they might be swamped by rising sea levels over the next 150 years.

The dangerous and polluting reprocessing of spent fuel at Sellafield should be stopped, but if acceptable means of long-term waste storage can be agreed, a new generation of nuclear power stations may be built in the UK, probably on the sites of existing stations.

Unconventional oil and gas

There are large reserves of 'unconventional' oil and gas (see box on p.55). The oil is in tar sands in the Canadian province of Alberta and elsewhere, and in oily shales in the Colorado region of the US and other areas. The gas is methane (natural gas) trapped in coal seams and in hydrates on the ocean floors and in the Arctic tundra.

There is already some experimental extraction of oil from the Canadian tar sands, with the expectation that this extraction will become commercially viable, as the price of oil rises later in the century. It is more costly to extract oil in this way than simply pumping it out of the ground, and requires a greater energy input. As for nuclear power (see audit on p.141) the total energy input must be subtracted from the energy produced, in calculating true cost.

The trapped methane is very difficult to access. In fact, the tundra and ocean-floor methane is dangerous, since it could be released into the atmosphere by global warming, acting as a greenhouse gas and spurring further warming in a positive feedback loop.

If the oil extracted from tar sands is used as a fuel for transport, this will add to carbon dioxide emissions and exacerbate the global warming problem. However, if this oil were used to produce hydrogen, and if in this process the carbon dioxide could be trapped in some way, rather than released into the atmophere, then these large reserves of unconventional oil could still be used. There is thus considerable interest today in the question of whether ways can be found to safely 'sequester' carbon dioxide for thousands of years or longer, in much the same way that natural gas has been sequestered in rock formations for millions of years.

One tonne of tar sand yields only 80 litres (0.055 tonnes) of oil. Extraction is very enviromentally destructive in the extraction area [6].

Sequestering carbon

Various ideas have been proposed to sequester carbon to control or reverse global warming. One suggestion is to use the deep oceans for storage [4]. Carbon dioxide piped under pressure to a depth of 3 km, or more, would settle as a pool on the sea bed.

The problem with this approach is that storing carbon dioxide in the deep oceans could prove a risky experiment, if carried out on a large scale. Upwelling ocean currents eventually bring deep waters back to the surface, and dissolved carbon dioxide makes the water more acid, which could affect the marine organisms on which the whole life of the seas depend.

Another idea is to spread iron salts on the ocean to encourage plankton growth. Plankton sink to the ocean floor when they die, carrying down carbon absorbed from the atmosphere during their growth or, if the plankton are eaten by fish, the carbon sinks to the bottom as fish bone. However, there might be considerable risk of disturbing natural cycles, if the spreading of nutrients were attempted on a large scale [5]. The whole rhythm of ocean life, the movement of fish, whales and birds from their feeding grounds to their breeding grounds is subtly interconnected with the spring bloom of phytoplankton in the Northern oceans and the bloom in the Southern hemisphere later in the year.

A sustainably managed forest is carbon neutral. New growth is balanced by timber removed.

One simple way to sequester carbon is by planting trees and maintaining existing forests. Trees absorb carbon dioxide from the atmosphere while they are growing, and this carbon remains sequestered for as long as the trees remain standing. When trees are felled, the carbon remains locked away for a while as paper, in wood furniture, or timber used in house construction. Eventually this paper and wood is burnt, or rots, and the carbon returns to the atmosphere. In mature forests the carbon absorbed by new growth is to some extent balanced by carbon released by decay of fallen trees, but mature forests in temperate zones continue to absorb carbon, if left standing, because the ground layer of humus increases in thickness year by year.

There is still controversy as to how effective forests will be in mitigating against global warming. Carbon is stored in the trees, but if climate change later causes forest die-off this carbon will be released.

Where large, new, permanent forests are planted, as China is planning to do in the middle and upper reaches of the Yangtze to control river flow and diminish risk of flooding, significant sequestering of carbon can be achieved. A halt to the destruction of remaining old-growth forests, and more planting of new, permanent forests will help towards control of global warming provided this warming does not cause forest die-off.

A limited amount of carbon dioxide could be sequestered in empty oil and gas fields and perhaps a vast amount could be stored in porous rock strata under the sea bed. The Norwegian company Statoil have started pumping carbon dioxide into a 200 metre thick layer of porous sandstone, the Utsira formation, 1000 metres under the North Sea, 240 km off the Norwegian coast (the pores were previously filled with sea water) . The porous layer is sandwiched between layers of impermeable rock so the carbon dioxide may remain trapped for a very long time, but this is a question still under investigation.

Carbon dioxide storage should be very safe in rock formations which have stored natural gas for millions of years.

Coal

Coal will remain abundant for some centuries after the conventional oil and gas are gone (see box on p.55) It will be still be used directly, perhaps, where a concentrated form of heat is needed, as for smelting metal ores and concrete production. These applications account for only 15% of coal used now in Britain. The main use of coal today (73%) is for electricity generation, in power stations where the coal is burnt, with release of carbon dioxide, sulphur dioxde and nitrous oxides into the atmosphere.

Alternatively, coal can be gasified. The 'town gas' used for city and home lighting in the early part of the last century (and used for cooking until it was replaced by natural gas from the North Sea) was 50% hydrogen. Gasification of coal is thus a well established technology (box on p.75). The final gas stream contains a mixture of hydrogen and carbon dioxide, but the carbon dioxide can be separated out. Coal can thus still be used in the future, as a source of hydrogen, or as a source of electricity derived from the hydrogen, provided pipelines are installed from the gasification plants to old oil or gas chambers, or to porous rock strata under the North Sea to sequester the carbon dioxide in an environmentally satisfactory way (Carbon Capture and Storage, CCS). Although it has yet to be established that carbon dioxide can be safely stored under the North Sea for thousands of years, exploring this possibility seems greatly preferable to continued dependence on nuclear power, as regards security against terrorist attack.

GASIFICATION OF COAL

If planning consent is given, Progressive Energy Ltd will begin construction in 2007 of one of the first UK coal gasification plants, at Onllwyn in South Wales. The coal will be gasified in cylindrical steel tanks around 2 m in diameter and 5 m high, at 1350 °C and at a pressure of 30 bar (30 times atmospheric pressure). Powdered coal, water and oxygen will be fed in from the top of these chambers, impurities will liquify and flow down to a water bath at the bottom where they will solidify into 'frit', a hard, glassy material. This frit could be safely buried but is, in fact, a useful construction material, as aggregate or formed into building blocks. Most of the lead, mercury and other heavy metals in the coal are trapped in the frit.

Some mercury remains, as vapour, in the gas stream but this can be removed with filters. The gas stream at this point is mostly carbon monoxide (CO) with quite a lot of hydrogen (H_2), some carbon dioxide (CO_2) and some hydrogen sulphide (H_2S). The gas stream now interacts with steam (one of the reactions described on p.54):

$$CO + H_2O = CO_2 + H_2$$

to produce a gas stream which is 55% H_2, 40% CO_2 and 5% H_2O, CO and H_2S. Bubbling through Selexol (a dimethyl ester of ethylene glycol) the H_2S and CO_2 are removed to leave a gas stream which is 80% hydrogen, with small amounts of CO, N_2 and CO_2. The H_2S and CO_2 can be sequentially recovered from the Selexol which is returned then to continue the process.

If the aim is electricity generation, the hydrogen fuels a gas turbine and the exhaust heat produces steam to drive a second turbine (Integrated Gasification Combined Cycle). The overall efficiency of electricity production is around 40%, which is a little above the efficiency of today's coal-burning power stations. Sulphur emissions from the gasification plant will be much lower than those from a coal-burning station of comparable generating capacity, even where this is fitted with sulphur removal equipment in the smoke stack. Nitrous oxide emissions from the gasification plant will be lower also. The Onllwyn plant will have three gasification tanks of the size described above, generating 450MW of electrical power, equivalent to 375 3MW off-shore wind turbines.

(Source: Progressive Energy Ltd)

A group of energy companies (BP, Shell and Scottish and Southern Energy) are proposing to build a 350 MW hydrogen-fired electricity generating station near Peterhead in Scotland (equivalent to 350 3MW wind turbines). The hydrogen will be reformed from North Sea natural gas and the carbon dioxide produced in this process will be piped to the Miller oil field, 240 km offshore, for enhanced oil recovery and carbon dioxide storage.

Carbon dioxide, pumped into an emptying oil field, mixes with the remaining oil to make it more fluid, allowing further extraction. It is estimated that this project will yield 40 million additional barrels of oil and sequester 1.3 million tonnes of carbon dioxide [9].

Later in this chapter, when we look in more detail at the energy crisis facing the UK over the next 20–30 years, we shall see that coal and CCS could play an important role over this period. In fact, it is possible that coal could meet all present and future UK energy needs for a century or more. If UK coal reserves are insufficient, or if mining in the UK is no longer economically viable, coal could be imported from Australia, South Africa and from those two countries with vast reserves: the US and Russia. However, coal should not be imported from countries where safety and working conditions are below UK standards.

Whether, or for how long, energy from this 'clean coal' technology will be competitive in price with energy from renewable sources remains to be seen. Energy is needed to make, maintain and operate the machinery used in mining, for transport to a port in the case of imported coal, for shipping, transport to the gasification plant, pulverisation, piping carbon dioxide to rigs in the North Sea and drilling down to suitable sandstone layers, when the storage capacity of old oil and gas fields has been fully exploited. At the present time the energy for all this comes, relatively cheaply, from oil and coal used without concern for carbon emissions. The total process will become more expensive later in the century, when clean energy must be used both for manufacture and maintenance of trucks, ships and all the other equipment involved, as well as

Where extraction rigs and piping are still in place, these can be used for carbon dioxide sequestration.

for transport. However, coal will remain a relatively low cost source of energy for as long as it's price remains low on the global market.

To make a comparison with, for example, offshore wind power, it would be necessary to estimate the energy needed to maintain and, from time to time, replace the turbines, once they were installed and to reposition them as necessary if their base or anchorage were shifted by tidal currents. Then we have to add the energy required to maintain the storage plants, or fossil fuel back-up, which would be needed to cover windless days. The wind, of course, is free.

Energy from waste

Britain produces each year 28 million tonnes of domestic waste. The UK level of recycling falls well below that of neighbouring countries and could certainly be brought to the 50% mark achieved across much of Europe (see p.79).

This would still leave 14 million tonnes to dispose of each year. If this waste is incinerated, rather than sent to land fill, the energy it contains can be used to generate electricity. But then there are problems of possible harmful emissions from the incinerator smoke-stack. In so far as the waste is paper, cardboard or wood, the energy generated is renewable energy (the carbon released during incineration balances carbon absorbed during the growth of the trees). But if the waste also contains plastic made from fossil fuels, this component adds to global warming. If the waste contains chlorinated plastics, such as PVC (poly-vinyl-chloride) incineration is environmentally disastrous, since the emissions from the incinerator stack will contain dioxins, which are highly carcinogenic.

Compact Power (see list of companies) have developed a very efficient waste gasification process.

For the future, the gasification of domestic waste may provide a much better solution than incineration. The gas stream can be heated to high temperature, to break down dioxins, and the chlorine can be trapped as hydrochloric acid. Heavy metals in the waste can be trapped in the slag which remains after gasification. The calorific value of domestic waste is in the range 2 to 4 MWh/tonne, so 14 million tonnes of this waste could contribute about 4% of UK electricity needs, and about twice this amount of energy as heat [7]. Incineration of waste contributes, at present, around 1% of electricity generation.

Energy can also be derived from straw, animal slurry, chicken litter and other farm waste, but we have to be cautious about generating energy in this way. Traditionally, on a mixed animal and arable farm, straw was rotted down with animal manure and spread back on the land as fertilizer. Today, in the UK, beef and dairy herds and arable crops are not only sited on different farms but, to a large extent, in different regions. We have thus, as Wendell Berry puts it, created two problems where there was no problem before. One is how to dispose of animal slurry and the second is the fact that arable land, fed with inorganic fertilizer, becomes low in organic matter, is more readily eroded, retains water less well in time of drought, and may become deficient in the soil micro-organisms which are essential for the mobilisation of minerals to produce a crop of high nutritive value. One of the aims of sustainable farming is to return organic matter to the land to produce a 'rich' soil which will produce healthy, nutritious crops, resistant to disease, and hence not requiring the application of pesticides.

For sustainable farming, then, if energy is to be derived from farm waste, this must be done in a way that retains the long-term fertility of the land. It is possible to digest slurry and sewage sludge to produce natural gas and a clean-smelling liquid fertilizer. Biodigesters are popular across rural China. The gas produced can be used for cooking, lighting, electricity generation or as a fuel for tractors, vans or other machinery.

The conversion of organic matter to natural gas is achieved by micro-organisms in the sludge.

At Holsworthy in North Devon, in a region of dairy farms, a biodigestion plant of this kind takes in, each day, 400 tonnes of cow slurry, pig and chicken manure together with waste from a local food processing plant. The natural gas produced in the digestor fuels an internal combustion engine driving an electric generator feeding 1.6 MW of power into the grid, 24 hours a day, 7 days a week (box on p.81).

**LEVELS OF RECYCLING
IN EUROPE**

(Data from *Independent
on Sunday*, 14 Sept '05)

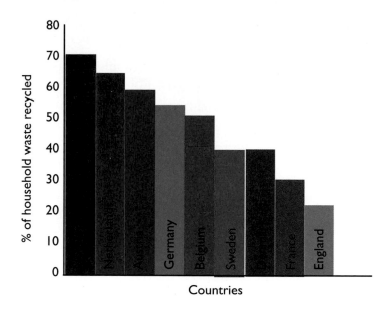

There is no reason, in principle, why human sewage should not be treated in a similar way. The only problem is that, in urban and industrial areas, heavy metals (zinc, cadmium, nickel and other metals used in many industrial processes) and other pollutants can contaminate sewage sludge, making fertilizer produced from such sewage unsuitable for spreading on the land. Sewage sludge biodigestion, at present, contributes 0.3% of electricity generation in the UK.

From the estimates given earlier in this section for domestic waste, and in the box on p.81 for farm waste, it seems that 7 to 10% of today's electricity needs could be met from these sources, if sewage contained no heavy metals. Even with this proviso, waste could only supply 2% of today's total energy need, which does not go far to meet the shortfall discussed in Chapter 3.

Biodigestion and gasification will provide good ways to treat waste, and could produce some useful energy, but cannot contribute significantly to total energy need.

The upcoming energy crisis (UK)

The most optimistic projections suggest that renewable energy sources could be supplying 10% of UK electricity need by 2010 and 20% by 2020 [8]. If this target is achieved, this will do no more than replace the power from the existing nuclear stations which will sequentially come to the end of their working lives during this time. This change will not contribute to any reduction in carbon dioxide emissions.

By 2020, the North Sea oil and gas fields will be exhausted, and deep mining of coal in the UK will no longer be economically viable [8].

A further problem will arise before 2020: emission limits set by EU legislation come into effect and will force modernisation or closure of the coal-fired power stations which today supply 29% of electricity [8]. A decision will have to be made as to whether to install emission reduction equipment, which will be expensive, or perhaps replace the coal-fired stations with gas-fired electricity generation. The second choice would leave the UK heavily dependent on imported gas for 80% of electricity generation and around 50% of total energy use. Most of the gas used in the UK, by 2020, will be coming through long pipelines from Russia, the Middle East and North Africa vulnerable to terrorist attack, political instability, or conflict of one kind or another, in the country of origin. Although security can be enhanced to some extent by importing some gas by sea as LNG (liquified natural gas) and by increased storage capacity to cover any short interruption of supply, it is not surprising that the UK Government is concerned about such heavy dependence on imported gas, and is considering building new nuclear power stations.

But, as we have seen, there is another alternative for electricity generation: to replace the coal-fired stations with coal gasification plants. The carbon dioxide from these plants could be piped to the emptying oil and gas fields of the North Sea, and this would enhance the recovery of remaining reserves [8]. Coal is more readily stored than gas, against possible emergency, and continued use of coal

There is also the possibility of capturing carbon dioxide from the smoke-stacks of existing coal- and gas-fired power stations for storage under the North Sea.

BIODIGESTION AT HOLSWORTHY

An incoming stream of animal slurry, poultry droppings and waste from food processing, is first pasteurized at 70°C for one hour (or can be brought to 95°C if there is reason to suspect the presence of pathogens resistant to treatment at 70°C). It is then digested for 3-4 weeks at 38°C. The electricity generator, and its exhaust, provide the heat needed for the pasteurization and digestion tanks, with the balance piped to Holsworthy, one and a half miles away, where it heats the hospital, the health centre, two schools, a sports hall, the council offices and 150 homes (at least, this is what it will do when this project is completed). 1.6 MW is not a great amount of electrical power (equivalent to 1.6 off-shore turbines) but the waste is pasteurised for return to the land and useful heat is generated.

The inflow of pasteurized slurry passes continuously into the digestion tanks and the outflow is an excellent fluid fertilizer, still containing about half the organic content of the inflow and milder, as a fertilizer, than the in-going poultry and pig manure. The lorries which bring in the slurry from the farms fill up with this fertilizer for the journey back to the farms, for its return to the fields. This is a win-win situation, in view of the need of modern animal and poultry farms to dispose of waste. The fertilizer from a biodigestion plant is well suited to grassland if not to arable land. To supply the Holsworthy plant about 8000 cows are needed, plus a comparable number of pigs and hens.

There are about 12 million cows in Britain, and a large number of pigs and hens. These could supply 1500 plants the size of the one at Holsworthy:

$$1500 \times 1.6 \times 24 \times 365 \text{ MWh} = 21 \text{ TWh per year,}$$

or about 6% of electricity need, plus community heating in the regions near to the digestion plants.

This may be an optimistic estimate of the amount of electricity which could be generated in this way. A report from British Biogen (Anaerobic Digestion of Farm and Food-processing Residues, 1997) gives a figure of 9.4 TWh/year rather than 21 TWh/year. Perhaps not all cows produce as much slurry as those fed on the lush pastures of North Devon. A more cautious estimate would be 3% to 6% for the possible contribution of biodigestion to electricity need.

(Source: Holsworthy Biogas Ltd)

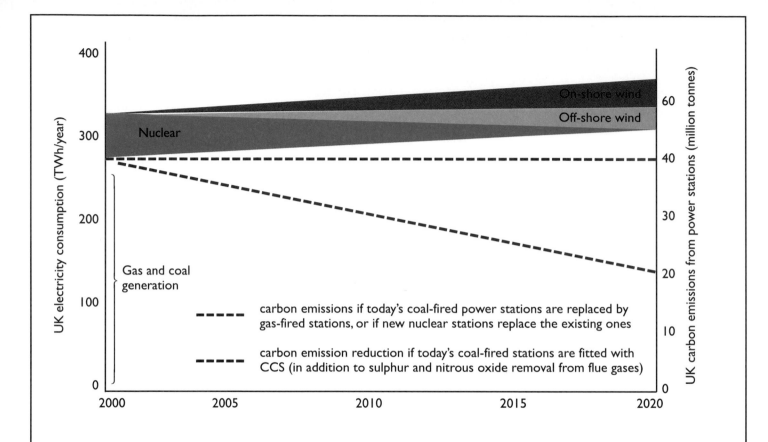

THE CASE FOR 'CLEAN COAL' RATHER THAN NUCLEAR OR GAS

These graphs show that if electricity consumption is allowed to rise, renewable sources will do little more than meet increased demand over the period to 2020. Replacing the existing nuclear power stations with new nuclear build to the current supply level, or replacing nuclear with gas-fired electricity generation, would only maintain carbon emissions at their present level. Carbon capture from the existing coal-fired power stations, or from coal-gasification plants of comparable generating capacity could markedly reduce the carbon emissions arising from electricity generation over this period.

would broaden the energy supply base. Nuclear power is suited to producing a steady electricity supply (base load). Coal is better suited to providing back-up for intermittent renewable sources. Also coal, with CCS, offers better hope of reducing carbon dioxide emissions by 2020 (see graphs on p.82). In the scenario for 2050 on pp.98–99 I have therefore made a firm choice for coal with CCS, rather than for nuclear. There may be problems to be overcome, and cost, in sealing the multi-punctured rock cap of an oil or gas field and monitoring to check for carbon dioxide escape, but these seem much less daunting than the problem of storing nuclear waste safely for 100,000 years.

By 2020, it is to be hoped that the cost of PV will be much reduced, so that, over the next period (2020–50) south-facing PV roofs and facades, and further expansion of wind and tidal current power, with perhaps some wave power, will contribute further electricity generating capacity. Over this period global oil production will fall. Cars, buses and lorries will be increasingly fuelled by hydrogen. If coal gasification plants have replaced today's coal-fired power stations, with sequestration of carbon dioxide under the North Sea, then, as renewable sources begin to play a greater role in electricity generation, these plants could now produce hydrogen to meet transport needs. There is also the possibility of increasing the area of woodland and setting large areas to willow coppice to produce wood chips for hydrogen generation.

Summary and conclusions

Continued reliance on nuclear power, and the building of a new generation of nuclear power stations in the UK, is a question which is going to be given serious consideration, but nuclear power is fundamentally dangerous or, more precisely, a nuclear accident, or terrorist attack, can lead to contamination of large areas of land for a very long time. Also, nuclear power for peace is but a step away from nuclear power for war.

Coal may continue to play an important role in energy generation in the UK, if the carbon dioxide produced when coal is gasified can be safely sequestered under the North Sea. However, when considering the 'clean coal' technology we have to bear in mind that safe carbon dioxide sequestration is untested in the long term. Energy from waste of all kinds, domestic refuse, slurry and human sewage could make a small but useful contribution to electricity and heating needs.

The chapter concludes with a timetable for UK energy development. The UK Government plans to increase the renewable contribution to *electricity* supply to 10% by 2010, 20% by 2020, and so on, but we have to remember that electricity represents only one fifth of total energy use.

CHAPTER 5

REDUCING TRANSPORT

Transport now accounts for some 30% of total energy use in Britain (see the pie chart on p.19) with almost all road, rail, sea and air transport dependent on fuels refined from crude oil. The availability and low cost of oil, even after the price hikes of the 1970s, has shaped the past 50 years: increase in air travel and car ownership, increase in the distances goods are transported. Energy statistics reflect day-to-day experience. Transport energy use has been rising steadily in the UK since 1960. This is a trend which will have to be arrested, or reversed, if any serious attempt is to be made to reduce carbon dioxide emissions.

Global oil reserves are expected to last some 30–40 years, and land transport, at least, can be adapted to run on natural gas, or hydrogen derived from natural gas. There is no immediate shortage of transport fuel. However, apart from the problem of global warming, it is prudent to look ahead to the time when global oil and natural gas supplies will no longer meet demand, and prices of these fuels will begin to rise irreversibly. Some decisions taken now will have long-term consequences. We have noted earlier the need for improved insulation in housing and the future potential value of planting now more forest and woodland. In this chapter we discuss ways in which the need for transport can be reduced in the UK, while maintaining, or enhancing, the quality of life. We consider the potential value of maintaining, or restoring, the structure of farming for greater local and national self-sufficiency, and of designing new urban areas with housing, work, schools and hospitals all within easy reach.

Transport of farm animals and food

Some transport today seems rather unnecessary. Caroline Lucas, a Member of the European Parliament, in her report 'Stopping the Great Food Swap' [1] brings together some amazing statistics. The UK exports each year to continental Europe:

> 33,100 tonnes of poultry meat
> 195,000 tonnes of pork
> 102,000 tonnes of lamb
> 100,000 live pigs
> 47 million kg of butter
> 111 million litres of milk

while importing from continental Europe each year:

> 61,400 tonnes of poultry meat
> 240,000 tonnes of pork
> 125,000 tonnes of lamb
> 200,000 live pigs
> 49 million kg of butter
> 173 million litres of milk [2].

It is difficult to see who benefits from this exchange, but not difficult to imagine the energy consumed, the number of lorries queuing up at the Channel tunnel and the Channel ports, nor the pollution they spew out on the way there. Refrigerated lorries, of course, use additional energy to run the cooling unit.

In another report [3], Angela Paxton provides ten case studies of wasteful energy use in our present system of food production and distribution, including a study by the Wuppertal Institute in Germany of a tub of strawberry yoghurt: strawberries from Poland, corn and wheat-flour from the Netherlands, jam from West Germany, the aluminium cover of the tub manufactured 300 km away from the yoghurt producer.

The supermarket system of food distribution is inefficient, from an energy viewpoint. Vegetables grown in a remote part of Britain are brought to a central packing station then distributed across the country, some returning to a supermarket near to where they were grown. UK supermarket chains now import fruit and vegetables not only from continental Europe but also from all over the world. Much of this produce is brought in by air.

At the present time, when oil is still abundant and the effects of global warming are not too evident, a system of food production and distribution has developed which is fundamentally unsustainable. It is pleasant, perhaps, to have a choice, in the supermarket, between butter from Ireland, Somerset, Jersey, the Alps and New Zealand, and a choice of other produce from all over the world, but a great deal of transport has been needed to achieve this result.

The scale of energy use for food production and distribution is studied in another report [4] which has been produced by Sustain in collaboration with the Elm Farm Research Centre. This report details the energy use involved in the transport by air of strawberries from California, blueberries from New Zealand, baby carrots from South Africa, etc. In the UK in 1998, the transport of food-related commodities (agricultural products, live animals, foodstuffs, animal fodder and fertilizer) amounted to around a third of all commodity movement by road. If the transport associated with the production and distribution of pesticides, packaging material, etc. is included, it is estimated that food production and distribution in the UK could account for up to 40% of all road freight [4]. This is a highly significant figure, in the context of climate change and the energy patterns for the future which we are studying in this book. It seems that carbon dioxide emissions could be significantly reduced in the UK, simply by return to a system of food production and distribution closer to the one which worked perfectly well 50 years ago! During World War 2, when food imports were severely restricted, nutritional levels for the very poor were higher than they are today. Over the past 50 years we have become accustomed to year-round availability of a wide range of fruits and vegetables, but it would certainly be possible, in principle, to return to local production of all the main food items, with imports added for variety.

The Angela Paxton report [3] also includes the diagram reproduced here as the box on p.89 showing how much animal feed is now imported into Europe from other parts of the world. Research at the Institute of Grassland and Environmental Research at Aberystwyth, in Wales, suggests imports of dairy cattle feed could be halved, with consequent halving of energy use, simply by growing more red clover (box on p.91). It may be argued that developing countries need to export to prosper, which is true, but surely a country should only export food and animal feed after all its own people are adequately fed. This point will be discussed further in the next chapter.

Also, the growth of soya beans in Brazil, for example, is leading to the destruction of the Amazon rainforest.

Farmers' markets, farm shops and local food production

The number of farmers' markets in Britain increased from zero in 1997 to over 400 in 2002. There are also now many farm shops and home-delivery schemes for locally-produced vegetables. These developments were perhaps mainly a result of the BSE (Bovine Spongiform Encephalopathy) crisis. People felt that food produced locally could be trusted. In this they are surely right. The vegetables at the farmers' market will usually have been harvested that morning. They will taste better, and will be more nutritious, than vegetables which have travelled a longer distance. The level of vitamin C in vegetables falls off rapidly from the moment they are cut. The produce in the farmers' market will also be free of all those preservatives which have been added to, or sprayed onto, supermarket food to prolong shelf life. Local production also reduces congestion on the roads and the air pollution caused by lorry and air transport.

At the farmers' market, or the farm shop, one can talk to the producer, ask if the apples or lettuces have been sprayed, or how free-range, really, are the hens whose eggs are on display. The problem with food coming from some distance away is not only the energy used in transport and the air pollution involved. The further away the food is produced, the less we know about how much pesticide has been used, the working conditions imposed on workers in the fields or the cruelty to animals involved.

Imported chicken is often 'bulked up' with water, and/or pork and beef proteins extracted from hides (*Guardian* 22 May '03).

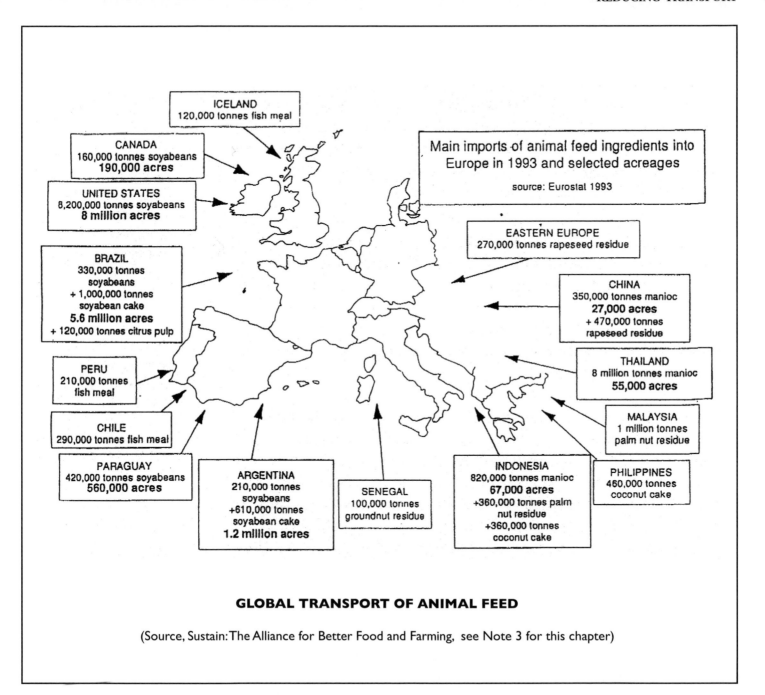

GLOBAL TRANSPORT OF ANIMAL FEED

(Source, Sustain: The Alliance for Better Food and Farming, see Note 3 for this chapter)

If the need to limit transport energy use leads to further expansion of local food production and distribution, this can be seen as a favourable development, reducing the international spread of disease, the cruelty of intensive poultry and animal production and transport of live animals. Farmers' markets, farm shops, and door-to-door delivery of vegetables, make high-quality food affordable, as well as allowing us to know more about how the food is produced.

There is a big demand today for ready-prepared meals. A range of such meals, made from local ingredients, can be provided at farm shops and farmers' markets.

Could the advantages of the farmers' market be combined with the convenience of supermarket shopping? Possibly an enterprising supermarket chain could decide to roof over an area near the main entry to their stores for a weekly farmers' market. This might prove profitable to the supermarket, in spite of competition from the farmers' market, because of increased turnover on market day. The Curry Report [5] published by the Policy Commission on the Future of Farming and Food in January 2002, following the foot and mouth crisis, proposes further Government support for local food production and distribution, as part of the solution to the current farming crisis, and suggests this could become 'mainstream within the next few years'. The report recommends that retailers (supermarkets) who give over a portion of their store as an outlet for local producers to sell directly to the public should receive business rate relief on that part of their premises.

We shall see in the next chapter that existing international agreements to limit carbon dioxide emissions, under the Kyoto Protocol, barely begin to address the problem of global warming. When the problem is eventually taken more seriously at the international level, more stringent limits will be placed on emissions and Governments will have to discourage the use of fossil fuels by increased taxation, and include aviation fuel within emission targets (which it is not at present). Together with steady increase in the price of oil, this will ultimately provide the solution to the current farming crisis in the UK. Local produce will become cheaper than produce brought from a distance. If food exports were diverted to home consumption, the UK could be 80–100% self-sufficient in most items, though relying heavily today on imported fruit and vegetables (box on p.93).

RESEARCH AT ABERYSTWYTH

GROWING different crops as cattle fodder would slash Britain's dependence on imported feed, say researchers in Wales.

Because of BSE, British farmers are banned from feeding meat and bone meal to cattle. So unlike pigs cattle shouldn't get foot and mouth through infected feed. But the ban has resulted in heavy reliance on imported vegetable protein to supplement feeds.

Over the winter, farmers feed their cattle silage, made by fermenting grass. Cows fed only grass-based silage produce 15 to 20 litres of milk per day. The yield can be increased dramatically, to 25 to 30 litres, by adding 6 to 8 kilograms of protein concentrate to the cows' daily rations. This means importing around 1 million tonnes of soya and 175 000 tonnes of maize gluten a year.

Roger Merry, Raymond Jones and Michael Theodorou at the Institute of Grassland and Environmental Research in Aberystwyth say farmers should make silage from red clover instead. "We could halve the imports of protein concentrates and could substantially reduce fertiliser application," says Jones. "We can produce safe food of consistent quality from home-grown crops."

The researchers have shown that cows fed half the normal amount of protein concentrate produce almost as much milk as those on a full ration if the silage they eat comes from red clover instead of grass. Of the home-grown legumes available, such as beans, peas, lupins and lucerne, red clover is the best bet because it will grow almost anywhere In Britain. "It's a very robust legume," says Jones.

Growing red clover in rotation with grass could also reduce the amount of silage required for the winter. Red clover can trap up to 300 kilograms of nitrogen per hectare per year, while its powerful roots reinvigorate soil trampled by grazing cattle. As well as reducing the need for fertiliser, this improves grass growth, allowing cattle to graze outside for up to two extra months each year. So the animals do better, as fresh grass is more nutritious than silage.   **Andy Coghlan**

(Source: *New Scientist*, 17 March '01)

There is an optimistic assumption here that climate change will not bring such constant cloud and rain to the UK that farm production is affected. Warmer temperatures at the equator will lead to increased evaporation from the oceans, and the Atlantic weather fronts move up towards Britain and Western Europe.

As the cost of energy rises, organic farming will become more competitive with intensive farming (that is, farming with heavy use of inorganic fertilizer and pesticides). For intensive farming, 0.4 kWh of energy is needed to produce 1 kg of cereal, while for organic farming only 0.072 kWh is needed [6]. Intensive milk production requires three and a half times the energy needed for organic production [4].

Increased tax on fossil fuels will be very unpopular unless balanced by corresponding tax reduction in some other area. Germany increased taxes on petrol, heating oil and natural gas in 1999 using the revenue to reduce employer and employee contributions to pension funds. Sweden increased taxes on diesel fuel, heating oil and electricity in 2001, while lowering income tax and social security contributions. The Netherlands introduced a general fuel tax as long ago as 1988, offset by decrease in income tax [7].

Cars, trains and planes

Any discussion of how to reduce energy use in transport must, of course, face up to the problem of the private car. I believe environmentalists lose credibility when they suggest that we must give up our cars and take to bicycles. Cycling is fine as a leisure activity, but most of us will find it difficult to relinquish the personal mobility in all weathers that the car provides. We shall, though, be prepared to drive smaller cars and use public transport in cities, and for inter-city travel.

In Chapter 1 we have seen how Susan Roaf uses a small electric car for local journeys recharged from the PV panels on her roof. The problem with electric cars today is that batteries are heavy, have to be replaced every few years, contain lead, or nickel and cadmium (which are harmful if released into the environment) and store only enough energy to give the electric car a limited range. Further research, and perhaps lithium-polymer batteries, may begin to solve some of these problems [8].

For those living in rural areas or living, like the Roaf family, in the suburbs of a town the size of Oxford, a small car of limited range is very adequate for driving to work, shopping, visiting friends locally, going to a film or driving to a pleasant spot for a country walk. In parking areas there could be 'coin in the slot' power points, so that electric cars could be recharged while parked.

For those living in larger towns and cities, the future lies in better city design so that, for day to day needs, the use of a car is neither necessary nor as convenient

UK SELF-SUFFICIENCY IN FOOD (1998)

(Source, Sustain: The Alliance for Better Food and Farming, see Note 4 for this chapter)

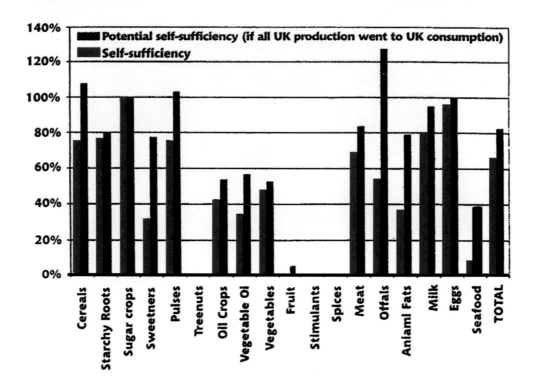

as travel by light rail or bus. There must be better integration of residential, work, shopping and recreation areas all accessible by foot or by frequent bus and rail services.

For longer journeys, a reliable, punctual, fast and uncrowded train is a pleasant form of travel, in many ways more enjoyable than driving along a motorway. There is a clear need for an efficient rail system in the UK, for both passenger and freight transport, whatever this may require in reorganisation and Government subsidy. There could be arrangements for train travellers to order a hire car for pick-up at their destination, as passengers do now when flying. It would then be possible to drive in a small car to the nearest railway station and take the train to one's destination, where the hire car would be waiting. Ideally, it should be possible to check in luggage near the car park and collect it at some point near the car hire compound.

Inter-city rail travel consumes about the same energy (per passenger mile) as a double-occupancy car journey, one half, if compared with a single-occupancy car journey (see: *How we can Save the Planet*, Mayer Hillman, Penguin, 2004).

THE VOLVO 3CC

A prototype electric car with lithium-ion batteries and a range of 180 miles (290 km) between charging stops

(Source: Volvo UK)

No doubt many people will still like to own a hydrogen-powered car for weekend and holiday use, even if this is not used much from day to day. Others may join a car share scheme, to have a car available when needed. Families with children need their own car for longer journeys, as do those frequently using the motorways. Hydrogen-powered cars may be more expensive to run than the cars of today, but there will be the compensation that road traffic will be much reduced, so that driving on the motorway will be more pleasurable.

It is not difficult to see reduced energy use in transport as a move towards a safer and more pleasant way of life. General use of smaller cars could lead to dramatic reduction in road deaths, particularly of children. Behind the wheel of a powerful car, there is a great temptation towards speed and rapid acceleration. Behind the wheel of a small car one is in a different mood, even before turning the ignition key. Cities of the future could be designed not only for greater integration of residential areas and areas of employment, but also to allow children to walk safely to school. Modern light industry is much cleaner than industry in the past. An industrial park can now be surrounded by residential areas and, in general, work areas and residential areas can intermingle to reduce travel distances. There

HOUSE DESIGN FOR ENERGY EFFICIENCY

Extending the roof, as shown here, increases the PV area and shades windows when the sun is high in summer.

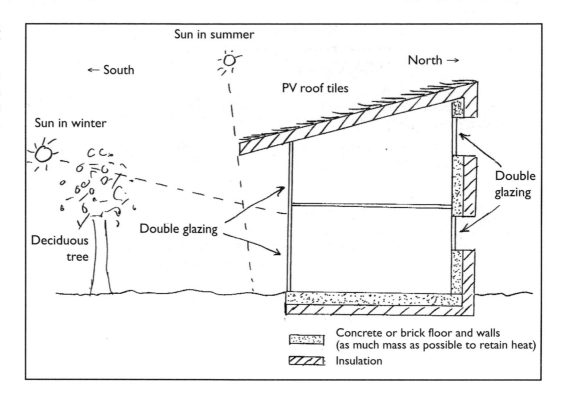

Sun in summer

← South

North →

PV roof tiles

Sun in winter

Double glazing

Double glazing

Deciduous tree

Concrete or brick floor and walls (as much mass as possible to retain heat)

Insulation

is also the question of whether everybody has to commute to work each day. For many people, in the future, teleworking from home might be possible for most days of the week.

We live at present at dead centre of the brief period in which the human race burns up the earth's store of fossil fuels, at the peak of the needle on p.5. The exhaustion of the readily-available fossil fuels will cause a radical re-think of where we are going, even if change has not been brought about already in response to the effects of global warming. Ultimately, human life depends on the productivity of the natural world. Only briefly have we been able to escape this fundamental law, as we use up the fossil fuels. We shall be forced to return to greater respect for, and dependence on, the natural world.

It is difficult to envisage the future for air travel. Flying will almost certainly become more expensive, since aviation fuel is at present neither taxed nor included in emission-reduction targets.

It would be foolish to believe that transition to an economy based on reduced transport will be easy to achieve. At the present time, transport, oil exploration and large-scale intensive farming are all still heavily subsidised within the EU. Considerable vested interest must be overcome to achieve change. New jobs must be created in renewable energy development and in redesign and renewal of city centres. But surely, no one believes that Britain today, with its clogged motorways and broken-down railways, is the ultimate in human social organisation.

A report from the Royal Commission on Environmental Pollution, 'The Environmental Effects of Civil Aircraft in Flight', published in Nov '02, recommends that no new airports, terminals or runways should be built, because air travel makes such a significant contribution to global warming, not only through carbon dioxide emission, but also through release of nitrous oxides in the upper atmosphere.

Scenario for the UK in 2050

We can now sketch out (in a speculative way!) a scenario for life in the UK around midcentury:

Houses in rural and suburban areas will incorporate the design features discussed in Chapter 2 with the additional consideration that, in the future, keeping cool in summer may be as important as keeping warm in winter. A roof overhang can increase the PV area and shade windows in high summer (see sketch on p.95). There will be efficient vacuum tube solar water heaters on the roofs. Supplementary space and water heating in winter will come, for the most part, from small CHP units fuelled by wood pellets made from waste wood. There may be a hydrogen-powered car in the garage, for use at weekends and for holidays with perhaps an electric car in the driveway for daily use, plugged into the mains at night for recharging.

City housing, as well as schools and hospitals, will be built to high standards of insulation and designed to make maximum use of PV and passive solar heating, as in the housing development at Beddington (pp.34–35). Most city dwellers, when leaving the city, will travel by train to their destination and have a hire car waiting there. Others will join a car share scheme, to have a car available at weekends and for holidays. Daily travel, to work or school, will be mostly by public transport.

Offshore wind power will supply electricity for the London Underground, the electrified part of the rail network, light rail in other cities and the needs of industry.

Inter-city trains may still run on diesel fuel, derived from seed oil crops, but buses will run on hydrogen for clean city air.

Farmers will grow enough oil seed crops to supply their farm machinery and to transport their produce to local markets.

Carbon credits will apply while the woodland is being established, though not later when they are being harvested sustainably.

During the early part of the 21st century, hydrogen will have been produced from natural gas, but later in the century grants will be available to encourage farmers and landowners to reforest the upland valleys of England, Scotland and Wales to control the constant flooding of York, Worcester and other riverside cities, and to gain credits under the Kyoto Protocol, or whatever international treaty may eventually replace the present protocol. The rising cost of imported timber will make this development economically attractive, and wood chips will be increasingly used for hydrogen production. By 2050, there will be woodland areas for recreation near all large cities.

Forestry work will be very popular with people who wish to drop out of mainstream society, the foresters living in 'ecovillages' of timber and strawbale houses in glades in the woodland.

Real-time telecommunication will be much more advanced than it is today, possibly with three-dimensional display screens, so that it will seem as though the person one is talking to is sitting right there on the other side of the desk, thus reducing to some extent the need to commute each day to work.

ENERGY BALANCE FOR 2050 SCENARIO

ENERGY PRODUCTION (per year)			TWh
WIND	7000 1.5 MW turbines on land (load factor 27%) (or, for each 1.5 MW turbine, 1000 1.5 kW roof-top turbines)		25
(off-shore load factor assumed to be 38%)	5000 3 MW turbines round the coast		50
	4500 5 MW turbines out in the North Sea		75
TIDAL CURRENT AND WAVE	approx. 1/4 of electricity (see p.28)		80
PV	approx. 1/4 of electricity (see p.28)		80
SEED OIL	1/10 of agricultural land (1.8 million hectares) set to rape, see Note 6, Chapter 1		20
DRY WOOD, FARM AND CITY WASTE	fuelling CHP plants (see Note 12, Chapter 5)	Electricity Heat	50 100
WOOD CHIPS (used to generate hydrogen)	clearance from 8.5 million hectares of woodland and coppice, with wood waste from parks, gardens and orchards (see Note 7, Chapter 3)		130
IMPORTED COAL (used to generate hydrogen)	50 million tonnes/year (the amount of coal used today) gasified, with carbon dioxide sequestration under the North Sea		330
		TOTAL	**940**

ENERGY USE (per year)   TWh

ELECTRICITY   Increased from today's consumption.   360
                               Savings can be achieved through
(today's use 329 TWh,   greater use of energy-saving light
see Note 3, Chapter 1)   bulbs and household appliances, but
                               this will be more than balanced by
                               increase in consumption due to the
                               greater number of homes and the
                               rural and suburban use of electric
                               cars, also perhaps more city trams.

TRANSPORT   Less than half today's use, due mainly   200
(aside from electrical)   to fuel cells being more efficient than
                               internal combustion engines, but also
(today's use 555 TWh,   greater use of rail, greater dependence
see Note 6, Chapter 3)   on local food production, cities designed
                               for walking and public transport, use of
                               electric cars for local journeys.

SPACE HEATING   Reduced to a third of today's use   150
                               by better insulation and increased
(today's use 463 TWh)   use of passive solar heating.

INDUSTRY and   Reduced to half today's use by   230
SERVICES   increased efficiency (see Chapter 3).
                               This is energy use in addition to
(today's use 463 TWh,   industrial and service use of
see Note 9, Chap. 5)   electricity and space heating,
                               which are included in sections above.

                                                      TOTAL   940

The energy balance for this scenario is shown in the box on pp.98–99. Total energy use (940 TWh) is half the energy used today (1860 TWh, see Note 3 for Chapter 1). This box is, essentially, a summary of all the ideas discussed in this book, and all the sums in the Notes.

1 TWh = 1000 million kWh

Some assumptions have to be made if we are looking this far ahead, and there is a basic assumption here that the UK population in 2050 will be much the same as it is today. Commercial and leisure activity remain at today's level, while the energy used for transport is less than half of today's use. Food and timber imports are much reduced but the economy depends on significant import of coal.

Government projections anticipate an increase in the UK population to around 65 million by 2030, followed by stabilization, or fall in numbers by 2050 [10].

In this broad-brush scenario, it is assumed that electricity can be stored (see Chapter 1) or that electricity and hydrogen are inter-convertible. All energy losses in storage, conversion, or transmission are ignored. In real life, rather greater energy production, or saving, would be needed to cover these losses.

This is only a 'scenario'. The reader is free to develop her/his own scenario, within the constraint that energy production and use must balance. There could be, for example, less use of coal with greater development of North Sea wind power.

The energy produced in this scenario from renewable sources is 610 TWh/year, that is about one third of the final energy used today. The balance of 330 TWh/year is supposed to be supplied by imported coal with sequestration of carbon dioxide under the North Sea. If this proves unsafe then, of course, coal cannot be used and this balance will have to be met by continued use of imported natural gas and/or by nuclear power.

After 2050 there can perhaps be further improvement in the efficiency of energy use to reduce total need to match renewable energy supply. Then neither coal, natural gas nor nuclear power will be needed.

The contributions from the various renewable energy sources, in the scenario presented here, are quite similar to those of scenario 1 in the RCEP 22nd report [11].

In the next chapter we shall consider an issue which has come up briefly in this chapter and in Chapter 3: whether the developed countries will be prepared to share more fairly the resources of the world, and the extent to which the development of renewable energy and the exhaustion of readily available fossil fuels may help to achieve a fairer global distribution of wealth. The wealthy today in Sao Paulo in Brazil, travel by helicopter, from home to work, or from home to a shopping centre, above the sprawling mass of humanity below. Under the present conditions of international finance and trade, the industrial nations could encourage the planting of sufficient seed oil and other fuel crops in countries of the third world to maintain and expand international aviation, at the expense of lower levels of nutrition in these countries. The wealthy could then jet around the world above the sprawling mass of humanity below. But a world divided into the very rich and the very poor is not an attractive future.

Summary and conclusions

The concept of redesigning the economy for reduced energy use is so attractive that this seems a sensible way to go forward, irrespective of the problems of global warming or the ultimate need to survive without the fossil fuels.

Locally-produced food is becoming increasingly popular and will become cheaper than food brought from a distance as further taxation is introduced to discourage fossil fuel use and as the price of oil rises.

It is difficult to beat a fast and punctual inter-city train service, combined with a small car to get to the station, and a car conveniently booked for hire at the destination station. This is done today for air travel and could be done for travel by train, making stations more like the reception and departure areas at airports.

No one enjoys a long and tedious commute to work. Teleworking and better integration of work and residential areas in cities could much reduce city transport.

The various estimates made throughout this book, brought together in the scenario on pp.98–99, suggest that renewable energy could be supplying some 600 TWh/year by mid-century, that is about one third of total final energy use today. If total energy use could be reduced to one third of today's level then renewable sources could meet future need. This may become possible as rising fuel prices force a realignment of the economy towards reduced dependence on transport and greater energy-saving. However, I have regarded it as more probable that final energy use could perhaps be halved by 2050, with significant contribution from 'clean coal'. Carbon dioxide emissions have been reduced to zero in this scenario (except perhaps for some direct industrial use of coal not included here). Hydrogen, biofuels, and electricity from sustainable sources provide the main energy for all construction, manufacture and transport. I believe the possibility of harnessing the power of nuclear fusion is still in the realm of fantasy, except in so far as energy produced in this way comes to us as sunlight.

It is not a bad future, except for the need to curtail air travel. We shall make fewer long-haul flights in the future, but value more highly each holiday at a distant location. The holiday will, in fact, be a richer experience, since the place we visit will have regained local character as the period of mass tourism comes to an end. Those countries which depend today on mass tourism for economic survival will have to adjust to this new reality.

The driving force towards the scenario of pp.98–99 will be the spiky, but inexorable, rise in the price of oil after the peak in global oil production is reached, perhaps by 2020 or sooner. Since natural gas can substitute for oil for many energy needs, the price of natural gas will rise with the price of oil. The UK economy must gradually adjust to higher energy costs and reduced, but more efficient, use of energy.

CHAPTER 6

COLLISION COURSE

In their report 'Collision Course' [1], Andrew Simms and co-authors consider the consequences which follow from the 70% increase in transport of internationally traded goods which has occurred over the twelve year period 1992 to 2004. Air travel has increased by 50% over this period and the number of cars on the world's roads has risen by 40% (from under 600 million to over 800 million). We can now project beyond 2004 and see the same trend continuing. As Simms *et al.* point out, the growth in emissions of carbon dioxide arising from increased global trade alone will overwhelm the reductions proposed at Kyoto in 1997, even if this agreement were signed and implemented in full. The UK provides a direct example of how this works: wind farms are being planned to generate renewable energy, but there are also plans for widening motorways and extending airports. This is considered essential for the future development of commerce, but will lead to increased consumption of fossil fuels. Aviation and marine fuels are not yet taxed, nor included in national targets for emission reduction. Global emissions may still be increasing in 2020 and the downward trend required to control climate change (the dashed line of the curve on p.4) may not even have started. The current growth of international trade and travel puts us on collision course with environmental constraints. It is this potential collision that we look at in more detail in this chapter, to try to see how catastrophe can be averted.

The Kyoto Protocol

If a solution is to be found to the problem of global warming this must clearly be achieved by international agreement. There is little point in any one country reducing fossil fuel use, if large areas of the world are still increasing their carbon dioxide emissions. Attempts to frame an international agreement to limit emissions began in Kyoto in '97, achieved a measure of success in Bonn in '01 and are still continuing, although the main contributor to these emissions, the US, has pulled out of the negotiations. The original aim at Kyoto was for each industrial country to agree to set a reduction target, say 5% below that country's emissions in 1990, to be achieved by 2012. It was clear from the start that some countries, notably the US (which was still at that time taking part in the negotiations) would have great difficulty in achieving any reduction, while others, notably the countries of the former Soviet Union, would have no problem at all, since the collapse of their economies had already brought their emissions below 1990 levels. This gave rise to the idea of carbon trading. A country emitting less than its quota might be allowed to sell carbon emission permits to a country prepared to buy such permits to allow it to exceed its quota.

Environmentalists are not happy with carbon trading, since they wish to see genuine reductions in emissions. However, such trading could provide an incentive for environmental conservation. A country conserving its forests or setting up a largescale reforestation programme would gain carbon credits.

Although the US has pulled out of the Kyoto negotiations progress has still been made. 125 countries have ratified the Protocol established in Bonn in '01 and the main industrial nations (except for the US) have accepted emission reduction targets, with penalties for failure to meet these targets by 2012. Europe is much more conscious than the US of the problem of global warming. There has been more flooding in Britain than usual in recent years, and severe flooding in Central Europe.

Future emission-reduction negotiations

It is accepted by all that the present Kyoto Protocol is still only a start. By 2012 new targets must be set to achieve genuine reductions in emissions. A musician, Aubrey Meyer, concerned about the problems of implementing the present agreement, has proposed a fair solution to the global warming problem (illustrated by the curves on p.106). The basic idea is to first set, by international agreement, a target level at which it might be feasible to stabilise the atmospheric carbon dioxide level and which it would be dangerous to exceed. This target might later have to be revised downwards, if evidence grew that serious consequences would follow from allowing the level to rise this much, or later relaxed if the effects of rising carbon dioxide levels on climate prove to have been exaggerated.

The curve on p.4 in the Introduction to this book is a tracing of the upper curve on p.106. As discussed in the Introduction, this curve shows the global emission reduction which would have to be achieved to stabilise the atmospheric carbon dioxide level at 450 parts per million, 60% above the pre-industrial level. Aubrey Meyer calculates what emission quotas are allowable, for each country, if this stabilisation is to be achieved, based on the principle of equity and justice: that everyone on earth should be entitled to equal rights in their share of carbon emissions. The more populous countries, such as China and India, would then be set emission quotas well above their 1990 emissions, allowing them to continue to increase their use of fossil fuels, as might be necessary for their development, and giving them the opportunity to sell permits. The industrial countries would be set targets which required them to reduce emissions or buy permits. Aubrey Meyer's aim is that by about 2030 emissions for each country would be matched to its population, within the total allowable at that time. After that, targets must be made more stringent, as all countries turn more to renewable energy, reducing carbon emissions substantially by 2050.

The wealth of the world today is largely concentrated in North America, Western Europe and Japan (map on p.107). These are precisely the areas where the fossil fuels have been used so intensively, up till now, creating the problem of global warming. It is only right that some of this wealth should be directed towards solution of the problem.

It will be convenient, in this chapter (though not ideal) to use the terms

---

Aubrey Meyer studied music at the University of Cape Town, plays the viola, has written ballet scores. He feels, and surely he is right, that his proposed solution to the climate change challenge has the 'harmony and internal consistency' of good music.

This will require some form of individual carbon rationing (see www.capandshare.com and *How we can Save the Planet*, Mayer Hillman, Penguin, 2004).

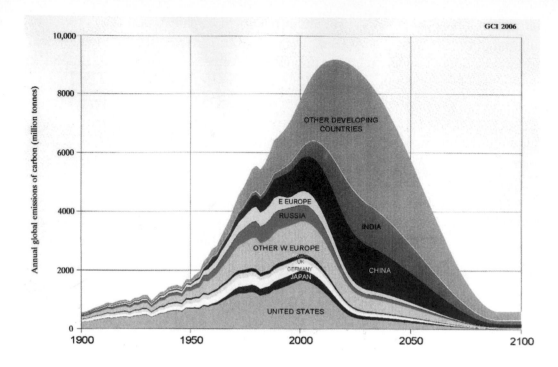

THE FAIR WAY TO REDUCE CARBON DIOXIDE EMISSIONS

(Source: Redrawn from *Contraction and Convergence*, Aubrey Meyer, Green Books, 2000. Copyright Aubrey Meyer, Global Commons Institute)

'developed' for these wealthy countries of North America, Western Europe, and Japan, including also Australia and New Zealand. The 'developing' world, as we are here defining it, then includes everybody else: Russia, Eastern Europe, China, India, South-East Asia, Central and South America and all of Africa. This developing world is sometimes now referred to as the 'South' or, with greater justification, as the 'majority' world.

At present, developing countries are not included in the Kyoto Protocol with its targets related to emissions in 1990. For most developing countries emissions were low at that time. It is accepted that they should be allowed increased emissions, but there is no agreement on how their targets should be set. It may be that only by general acceptance of per capita targets that a global agreement can be reached including both developed and developing nations. If this could be

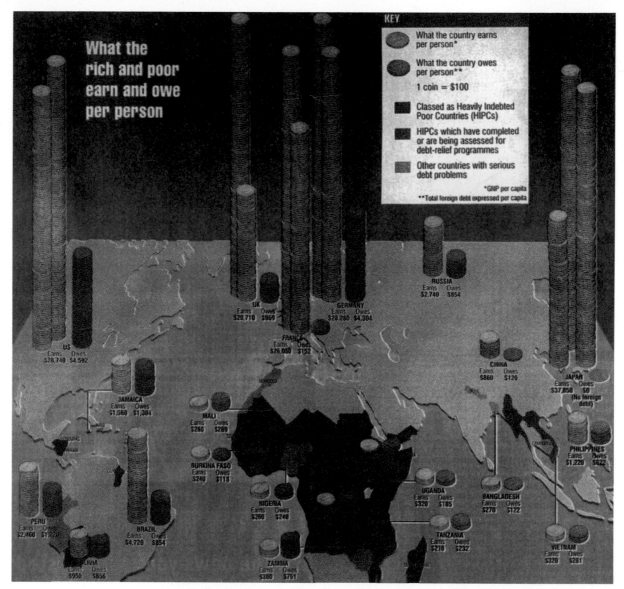

THE UNEVEN GLOBAL DISTRIBUTION OF WEALTH
(Source: World Development Report '98–'99. Copyright: Independent Newspapers Ltd)

Grey counters = GNP/capita   Red counters = Foreign debt/capita   Each counter = US$100
Figures for 2004 are given in Note 15 for this chapter.

achieved it would be of great benefit to the developing nations. They would have carbon permits to sell and there would be a flow of wealth from the developed to the developing world, distributing wealth more fairly.

It is perhaps not surprising that the US pulled out of the Kyoto negotiations. If North American emissions continue to rise, as they are still at the time of writing, signing up to any emission reduction treaty will commit the US to very large payments to countries with carbon permits to sell. However, the US cannot remain isolated from the rest of the world and will surely join emission reduction negotiations, in some measure, as the effects of global warming become more undeniable. The US may surprise us all by developing hydrogen-powered cars, huge wind farms, PV power, reforming natural gas, sequestering carbon, dramatically reducing emissions by mid-century, even without signing up to international treaties. This change towards greater reliance on renewable energy may be driven less by the effects of global warming, more by the need to reduce dependence on Middle East oil. The US is in a very favourable position, with abundant desert areas in which to develop wind and PV power. Also, about half the food grown in the US today is exported, so that with some reduction in food exports, there is abundant land on which crops could be grown to supply biofuels and bioplastics.

In a book listing '101 solutions to the problem of global climate change', Guy Dauncey and Patrick Mazza detail the changes needed to reduce North American energy consumption by a factor of four [2]. They suggest that improved efficiency in energy use could reduce energy need by a factor of two (fuel-efficient cars, low energy light bulbs and household appliances, improved industrial processes) and that a further factor two could be achieved by change in life style (redesign of cities for walking and cycling, eating more locally-grown food, travelling less, living more simply).

Given a factor-four reduction in energy use in the developed world, Dauncey and Mazza calculate the amount of renewable energy which would be needed, globally, to allow 7.5 billion people to live this simplified life style. They suggest

This wealth transfer is already happening in a tiny voluntary way when air travellers 'off-set' the carbon emission of their flight by contributing to 'carbon saving' in a developing country.

that, by determined effort, sufficient renewable energy could be on line by 2025 to achieve this aim (45% from wind, 20% from PV, 15% from biomass and biofuels, 15% from geothermal and 5% from tidal, wave and hydroelectric power). Their time scale is optimistic (a thousand-fold increase in wind power is required above today's level) but this is an inspiring book full of useful information (www.earthfuture.com/stormyweather).

For developing countries, the installation of wind turbines and PV, as well as small-scale hydro-electric generators and biogas digestors, could allow self-sufficiency in energy production, eliminating costly fuel imports.

Reserves of oil and natural gas

Behind the problem of global warming lies the problem of the exhaustion of the conventional oil and natural gas. It is notoriously difficult to foretell the time when the oil and gas will all be gone, since new oil and gas fields are still being discovered and many existing fields have not yet been exploited to their full potential. But we have seen, in Chapter 3, that many analysts now expect global oil production to peak before 2020. After peak production the price of oil must inevitably rise. Since natural gas can substitute for oil in many energy applications, an increase in the price of oil leads to rising gas prices. These increases in the cost of energy may lead to serious economic problems, both nationally and globally, but from an environmental point of view this is precisely what is required to bring global fossil fuel consumption sliding down the dashed curve of p.4, to control global warming, whether the Kyoto Protocol is implemented or ignored.

It would be possible, in fact, to use up all the known reserves of conventional oil and gas and still achieve the target of stabilising the atmospheric carbon dioxide level at 450 ppm [3]. International agreement on emission reduction will still be needed to ensure that the direct burning of coal for electricity generation is phased out, and that future use of coal and tar sands is directed to hydrogen production and sequestration of carbon dioxide (now often abbreviated to CCS, carbon capture and storage). Agreement on emission reduction and carbon trading is also needed to encourage the preservation of forest areas, and the replanting of forests which is so necessary to reduce soil erosion, control river flow and carry rain to arid areas.

Reduced supplies of conventional oil and gas will dramatically drive forward the spread of wind turbines and PV, already a growth industry today. If it proves possible to generate as much energy from renewable sources and CCS technology as is derived today from conventional oil and gas, and then continue to increase the available clean energy to meet predicted future needs for expansion of international trade and global growth in air and car travel, then the problem of the 'collision course', discussed in the first section of this chapter, will be solved.

However, the detailed study for the UK in this book suggests, to me at least, that it will just not be practicable to generate this much energy renewably and cleanly or, at least, that the cost will be high so that energy will become much more expensive. There is ample wind and sunshine but, as noted right at the start of this book, the energy which can be derived from these sources is diffuse and must be collected over a wide area. To produce hydrogen in any quantity from wind or solar power, large areas of turbines or PV panels are needed, power lines to water and electrolysis plants. All over the world there will be enough local renewable energy for a simplified, comfortable lifestyle of the kind sketched out in the scenario for Britain in 2050 in Chapter 5. But there may not be enough for frequent air travel for everyone, for a huge increase in the number of private cars or for continued growth in international trade. In this view I am not alone. The pro-nuclear lobby see an opportunity for revival of their industry, on the grounds that renewable energy supplies will be inadequate to meet need. Others believe that we shall have to set large mirrors in space collecting sunlight and beaming energy to Earth.

Although it is possible to imagine a lifestyle in which there is reduced use of

For each country, and for each region, honest accounting will be needed to establish exactly how much carbon has been sequestered in trees, over a given period.

Britain is in a favourable position to move to CCS, with relatively large storage potential under the North Sea. Globally, this may only be a short-term solution. Bernt Kuemmel (ref. in Note 4, Chapter 4) estimates that carbon dioxide storage capacity in saline aquifers and depleted reservoirs, world wide, would only be sufficient for 45 years at today's level of emissions.

energy, this will be extremely difficult to achieve from a political and economic viewpoint. For every country today, at least for every developed country, economic conditions dictate that there must be annual growth of GNP to maintain prosperity (GNP, the Gross National Product, is the sum total of all monetary transactions in the country during any one year). The higher the growth in GNP, the happier politicians become. A low growth rate means increased unemployment, higher welfare payments (in the developed world), less tax revenue, less spending, with danger of a spiral of recession leading to deteriorating public services: health care, education, support for the disadvantaged and aged.

Growth appears essential. But each time we go out to buy a newspaper, or a mobile phone, or carry out home improvements, and thus contribute to growth of the GNP, energy is consumed. For the newspaper this is the energy used to cut down and pulp trees or to recycle old newspapers (which also takes energy). Similarly, the production of the mobile phone, or manufacture of the material for home improvement, has involved energy use. Growth can be decoupled from energy use, to some extent, by more efficient use of energy. The DTI Energy White Paper, Feb '03, shows that, in the UK, GNP grew by 100% between 1970 and 2000, with only 10% growth in energy consumption. To maintain growth while *reducing* energy use is more difficult.

Energy consumption in the UK fell, over the period 1970–1985 as the heavy industries were phased out, but since then has increased again each year. It is now back above the 1970 level, mainly as a result of a doubling of energy use for transport. Thus prosperity in the UK today is based on precisely the increase in transport which I am suggesting in Chapter 5 (in agreement with other environmentalists) will have to be reversed.

If energy use in the UK increases by 1% per year until about 2015, it will have to fall by 1% per year thereafter to achieve the scenario of pp.98–99. It should not be too difficult to achieve this steady improvement in the efficiency of energy use.

It is, perhaps, not a question of 'will have to be reversed' but rather 'will inevitably be reversed', over the next 50 years, as the amount of renewable energy available proves insufficient to fill the gap created by the declining availability of fossil fuels, as the cost of carbon sequestration is factored into continued use of coal, and as the cost of transport rises. This will remain true even if a decision is made to replace the existing nuclear power stations in the UK with a comparable number of new stations (see Chapter 4).

Richard Douthwaite faces up to these questions in *The Growth Illusion* (Green Books, 1992) arguing that 'economic growth and increased energy use are inseparable', since any gains achieved by using energy more efficiently are eventually lost if there is constant growth of GNP. He discusses how we might move towards a stable, zero-growth economy and is optimistic for the long term:

> 'The first casualty of a move to a lower energy world would be transport, the most energy-intensive sector of all. Higher transport costs would allow small producers to re-emerge to use local resources to supply local markets. This would, in turn, create greater local, regional and national autonomy, reversing the concentration of economic power that has taken place in the last century. What a flowering of cultures, communities and individuals there could be!'

The whole question of how 'growth' of happiness and well-being can be achieved, with reduced use of energy, will need to be given serious consideration in coming years, at both the national and global levels. In the remaining three sections of this chapter, I shall gather together some of the evidence which suggests that today's global economy, besides depending so much on oil for transport, is also not serving the interests of the developing world and is reducing food security in the least-developed countries. Global trade needs to reformed in a way that not only reduces energy use but also ensures poverty reduction and a fairer distribution of wealth.

Globalisation

The second half of the 20th century saw unprecedented growth in international trade, investment and communication, a process we now call 'globalisation'. This integration of the global economy has brought prosperity and higher living standards to the developed world and to those developing countries which have been able to establish an industrial base for the manufacture of cars, TVs, computers, mobile phones or domestic appliances. The less developed countries have been left behind. If we compare the opportunites open to a child in a developed country, with access to computers, both at home and at school, with those of a child at school in an African village, without either paper or pencils, it is clear that each new advance in technology widens this gap between rich and poor.

The present position is summed up in a United Nations report, Escaping the Poverty Trap [4]:

This quote provides a concise definition of 'globalisation'.

> 'Gobalisation, that is, the increasing flow of goods and resources across national boundaries and the emergence of a complementary set of organizational structures to manage these flows, is tightening the international poverty trap of commodity-dependent least-developed countries and intensifying the vulnerability of least-developed countries which have managed to diversify out of primary commodity exports into exports of manufactured goods and/or services.'

Most developing countries have built up huge debts while, at the same time, the value of the commodities they export (tea, coffee, copper, rubber, etc.) has fallen (box on p.114). As we shall see in the next section, this has led to extreme rural poverty in many parts of the world.

Soon after World War 2, at Geneva in 1947, international negotiations began to establish a General Agreement on Tariffs and Trade (GATT). The GATT negotiations, from the start, sought to base trade on market forces. Prosperity for all, it was supposed, would follow from 'free', or 'liberalised', trade, that is, the gradual removal of all subsidies (the payments a Government might make to support its own producers, distorting the free play of market forces) and removal of all tariffs (import and export duties) and other barriers to the free international

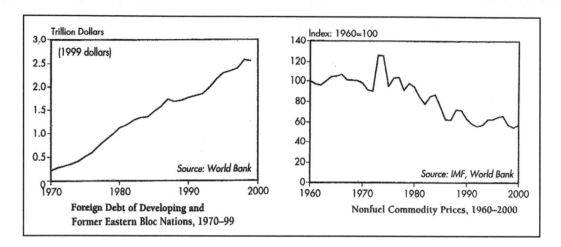

Foreign Debt of Developing and
Former Eastern Bloc Nations, 1970–99

Nonfuel Commodity Prices, 1960–2000

RISE OF THE DEVELOPING COUNTRIES DEBT AND FALL IN COMMODITY PRICES
(Source: Worldwatch Institute, Vital Signs, 2001–2002)

These trends have seen some reversal more recently, with a measure of debt cancellation and a rise in global demand for commodities (particularly from China).

movement of goods. Successive rounds of negotiation led, in '95, to the replacement of GATT by the World Trade Organisation (WTO) with 120 members at that time and further countries, notably China, joining since.

The main argument for the benefits of free international trade rests on the theory of 'comparative advantage' first clearly stated by David Ricardo in 1817. If every country produces only those goods it makes best, and then trades them for everything else it needs, everyone everywhere will have more of everything, because the resources involved will have been used more efficiently [5].

Possibly, in an ideal world in which all nations were at a comparable stage of industrial development, the free play of market forces might bring prosperity to all. In the real world, disparity in economic power has allowed the US and the EU to bias the WTO rules in their own favour. Subsidies are supposed to be scaled down under free trade, but the US and EU still maintain large agricultural subsidies, while insisting that developing nations lower tariffs on imported goods (box on p.116). This erodes food security in the developing country. The US and the EU also maintain substantial tariffs against goods, particularly processed goods, imported from developing countries.

Global trade based on rules biased in favour of the wealthy, is not bringing prosperity to all. One fifth of the world population today lives on less than $1 a day and a further fifth on less than $2 a day. Money is not everything, and perhaps one can be happy on $1 a day if one has land on which to grow food, a small surplus to sell in exchange for other necessities and access to clean water. But this is not the life which the poor live in many parts of the world. Roughly one billion people are malnourished and underweight and/or do not have access to clean water.

For thousands of years extreme disparity in wealth has been accepted as inevitable. But in the latter half of the last century extreme poverty began to be seen as intolerable. If it was possible to walk on the moon, to carry out delicate heart by-pass surgery and to sequence the human genome, why was it not possible to ensure that every child on earth was well fed and had access to clean water, basic education and health care?

The challenge for the new century is to eliminate extreme poverty, to develop a fairer system of global trade, to restore degraded agricultural land, to restore forests, to clean up the toxic chemicals now ubiquitous in the environment and the nuclear waste, and to do all this as the global reserves of oil and natural gas, on which today's economy depend, are nearing exhaustion. This challenge was summed up in the Introduction to this book in the phrase 'the problems which seem so overwhelming at the start of the new century'.

Finding solutions for the future may require a quite fundamental change in approach. Mention was made, in an earlier chapter, of the book *Factor Four* in which Amory and Hunter Lovins team up with Ernst von Weizsacker to show that life in the developed world could be just as comfortable as it is today using only a quarter of the resources used at present. The Lovins later collaborated with Paul Hawken to write *Natural Capitalism* (Earthscan, 1999) extending these ideas. As the title implies, their main point is that we must begin to value more highly the Earth's natural capital: the services provided free by nature, the soil, forests, rivers and seas on which all life depends. Natural capital is largely ignored by

---

THE UNDERCUTTING OF LOCAL RICE PRODUCTION

An extract from *Eyes of the Heart*, Common Courage Press, Monroe, Maine, USA, 2000,
by Jean-Bertrand Aristide, former president of Haiti

"What happens to poor countries when they embrace free trade? In Haiti in 1986 we imported just 7000 tons of rice, the main staple food of the country. The vast majority was grown in Haiti. In the late 1980s Haiti complied with free trade policies advocated by the international lending agencies and lifted tariffs on rice imports. Cheaper rice immediately flooded in from the United States where the rice industry is subsidized. In fact the liberalization of Haiti's market coincided with the 1985 Farm Bill in the United States which increased subsidies to the rice industry so that 40% of U.S. rice growers' profits came from the government by 1987. Haiti's peasant farmers could not possibly compete. By 1996 Haiti was importing 196,000 tons of foreign rice at the cost of $100 million a year. Haitian rice production became negligible. Once the dependence on foreign rice was complete, import prices began to rise, leaving Haiti's population, particularly the urban poor, completely at the whim of rising world grain prices. And the prices continue to rise.

What lesson do we learn? For poor countries free trade is not so free, or so fair. Haiti, under intense pressure from the international lending institutions, stopped protecting its domestic agriculture while subsidies to the U.S. rice industry increased. A hungry nation became hungrier."

Another example, from an article by Edward Goldsmith in the *Ecologist* (June '01 report)

The price of soya imported into India from the US would be $348 a ton instead of $155 if the US government did not subsidise it. No small farmer in India or elsewhere can compete with that.

---

economists, since it is difficult to give it precise money value, but perhaps, when the loss of natural capital is taken into account, the world was poorer in the year 2000 than it was in 1950, in spite of the growth of cities and apparent prosperity. Globalisation, so far, has been a very mixed blessing.

In the last section of this chapter I want to explore the changes that will be needed to move to more sustainable food production and distribution.

Global trade in food

In the 1970s, hybrid, high-yield varieties of maize, wheat and rice were developed at the International Research Centres in Mexico and the Philippines. In what has been termed the 'Green Revolution', these varieties have been widely planted across the world and have played a significant part in feeding a growing global population. But these varieties only give high yields on irrigated land, with ample input of fertilizer and heavy application of pesticides. Only a relatively wealthy farmer can afford these inputs and the purchase of hybrid seed. Poor farmers cannot compete and are forced to sell their land to a wealthy neighbour, work as hired labourers, move to the city, or move onto more marginal, less productive land. The best land is increasingly devoted to the production of high-value export crops (box on p.118).

The price of artificial fertilizer and pesticides is going to rise with any rise in the price of oil and natural gas.

In his book *Hungry for Trade* (Zed Books, 2000) John Madeley brings together material from 36 case studies presented at a conference organised by the Association of Church-related Development Organisations and an FAO (UN Food and Agricultural Organisation) study of food imports and exports for 16 developing nations. These studies show that trade 'liberalisation', that is, the forced lifting of tariffs restricting food imports, which Jean-Bertrand Aristide describes so graphically (box opposite), and the move to production of high-value crops for export, have devastated rural life in developing countries:

> 'In Ghana farmers cannot obtain an economic price for their produce, even in village markets. Their produce cannot compete with imported maize, rice, soya bean and chicken'

> 'In Kenya persistent food deficits and decreased incomes lead to families eating fewer meals per day'

> 'In Benim government incentives have led to increase in the land under cotton and increase in cotton exports but the food security of the poor has been undermined'

RURAL POVERTY

This extract is taken from an article by Peter Rosset in *Third World Resurgence*, no. 129/130, May/June 2001.

Around the world, the poorest of the poor are the landless in rural areas, followed closely by the land-poor, those whose poor-quality plots are too small to support a family. They make up the majority of the rural poor and hungry, and it is in rural areas where the worst poverty and hunger are found. The expansion of agricultural production for export, controlled by wealthy elites who own the best lands, continually displaces the poor to ever more marginal areas for farming. They are forced to fell forests located on poor soils, to farm thin, easily eroded soils on steep slopes, and to try to eke out a living on desert margins and in rainforests. As they fall deeper into poverty, and despite their comparatively good soil management practices, they are often accused of contributing to environmental degradation.

But the situation is often worse on the more favourable lands. The better soils are concentrated into large holdings used for mechanised, pesticide, and chemical fertiliser-intensive monocultural production for export.

Many of our planet's best soils — which had earlier been sustainably managed for millennia by pre-colonial traditional agriculturalists — are today being rapidly degraded, and in some cases abandoned completely, in the short-term pursuit of export profits and competition. The productive capacity of these soils is dropping rapidly due to soil compaction, erosion, waterlogging, and fertility loss, together with growing resistance of pests to pesticides and the loss of biodiversity.

The products harvested from these more fertile lands flow overwhelmingly toward consumers in wealthy countries. Impoverished local majorities cannot afford to buy what is grown, and because they are not a significant market, national elites essentially see local people as a labour source — a cost of proluction to be minimised by keeping wages down and busting unions. The overall result is a downward spiral of land degradation and deepening poverty in rural areas. Even urban problems have rural origins, as the poor must abandon the countryside in massive numbers, migrating to cities where only a lucky few make a living wage, while the majority languish in slums and shanty towns.

The reference for this section of the article is: *World Hunger, Twelve Myths* by Frances Moore Lappe, Joseph Collins and Peter Rosset with Luis Esparza, 2nd edition, Grove Press, New York, 1998.

In Bolivia, India, the Philippines, Honduras, Jamaica, Malawi, Mexico, Zimbabwe, Uruguay, all over the world the story is the same and disastrous for small farmers. There is a great deal of evidence now to show that global trade in food, the encouragement by the International Monetary Fund (IMF) of the export of food and commodities from developing countries to repay debt, and the 'dumping' of agricultural surplus from the US and the EU, that is, sale at prices below the cost of production, are leading to land degradation and increased poverty in rural areas of the developing world.

A country which tries to meet debt repayment by growing high-value cut flowers and mangetout beans, for export to the US and Europe, while relying on cheap grain imports to feed its population, is in a very vulnerable position. In a time of recession in the US or Europe the demand for cut flowers dries up. In years of poor harvest in the grain-producing countries, the price of grain rises on the world market. The main grain-exporting countries are relatively few: the US, Canada, France, Australia and Argentina for wheat; the US, Thailand, Vietnam and China for rice; the US and Argentina for maize. The world is becoming increasingly dependent on these countries for food security [6]. If the US seeks to reduce dependence on Middle East oil, an increasing acreage of agricultural land may be set to the production of biofuels and bioplastics and the US could cease to be a major food exporter.

In response to the need to feed a still growing world population, the large transnational seed producers are developing genetically-modifed (GM) crops which are resistant to herbicides or to pests, or suited to more saline soils. But there are problems with this approach: weeds and insects evolve, that is, in the course of random mutation new strains of weed or insect appear which are resistant to herbicides and pesticides currently in use, so that research directed to the development of new GM strains can never achieve any permanent solution and is expensive. GM seeds will never be affordable for poor farmers in developing countries unless the biotech corporations offer them at reduced price. But also, the wide use of GM seed is a serious threat to biodiversity. A corporation may offer, perhaps, a dozen different strains of rice, suited to different conditions. Traditional Indian farmers use tens of thousands of different rice strains which could be lost for ever if Indian farming moves to large monocrops of GM strains.

Low-input farming

In view of the problem of loss of food security associated with dependence on global trade in food, the energy needed for intensive farming, and the potential problems of the high-tech GM approach, some agronomists are suggesting that what is really needed is low-input farming. This approach depends on low-tech research, and exchange of information amongst farmers to increase yields, through simple water-retention methods on rain-fed lands, with planting of legumes to suppress weeds and to increase the organic content and nitrogen in the soil, with interplanting of complementary crops to reduce pest infestation. This is farming which drastically reduces costs of inputs, is affordable for the poor, and is more sustainable than intensive farming. As Jules Pretty and his colleagues have shown, substantial increases in yield can be achieved (box on p.122). In general, small-scale farming can give higher yields than intensive farming [8]. The box on p. 121 provides just a small example of what is possible. In Jiangsu province in China there has been rapid growth of fish, crab and shrimp farming in rice paddy fields, as a result of aquaculture projects launched in the mid '90s [9]. For aquaculture, pesticide use must be restricted, but rice yields increase, perhaps because of the fertilizing effect of the fish excrement.

Cuba provides another example of the potential for low-input farming. Up to 1990, Cuba's agricultural and food sector was heavily dependent on external support from the Soviet Union. Cuba imported 100% of wheat, 90% of beans, 57% of all calories consumed, 94% of fertiliser, 82% of pesticides, 97% of animal feed and was paid three times the world price for its sugar [7]. The collapse of the Soviet Union was thus catastrophic for Cuba in the short term. But Cuba turned to low-input, organic farming: crop rotation, green manuring, intercropping, and

SMALL-SCALE INTEGRATED FOOD PRODUCTION IN JAPAN

In this box I summarize an article by Mae-Wan Ho in the *Ecologist*, Vol. 29, No. 6, 1999. It is the story of two farmers in Japan, Tokao Furuno and his wife. They release ducklings into their rice paddy fields soon after the seedlings are planted. The ducklings eat up insect pests and the snails which attack rice plants, and they eat the weeds too. Duckweed is grown on the surface of the water, some eaten by the ducks, some harvested as cattle feed. Roach, in the paddy, feed on worms and duck faeces, and both fish and ducks provide manure for the rice. Towards the end of the growing season, the ducks are returned to their shed. Fed on waste grain, they lay eggs and mature for market. The eggs hatch out next seasons ducklings. The Furunos' two hectares yield annually 7 tonnes of rice, 300 ducks, some roach, and enough vegetables to supply 100 people (from the 0.6 hectare on which the Furunos grow organic vegetables). 'This kind of farming is not hard work, it is fun!' says Takao.

Mae-Wan Ho is now Director of the Institute of Science in Society and develops this theme more extensively in two articles: Dream Farm and Dream Farm II in the Institute's energy report, Which Energy?, 2006. Pretty and Hine [9] give other examples of fish and shrimp farming integrated in this way with rice production.

converted city gardens and plots to food production. Food intake was 2600 kcal/day/capita in 1990, fell to some 1000–1500 kcal/day during the transition and rose to around 2700 kcal/day by the end of the 1990s. Low-input farming, if it increases the organic content of the soil year on year, can also contribute significantly to carbon sequestration and could qualify for carbon credits under the Kyoto Protocol [10].

The present system of global trade in food is not sustainable. It is a system which could only have been established in a time of abundant oil. The intensive farming of rice in the US requires 15–25 times the energy needed to grow the same amount of rice in Haiti or India, where growing rice for local consumption requires little external input [10]. To the energy use of production in the US we have to add the energy used in global transportation. We saw in the last chapter that the transport of food to and fro across Europe will not be sustainable if serious efforts are made to tackle the problem of climate change. The same is true for transport of food across the world.

THE POTENTIAL FOR SUSTAINABLE FARMING

Jules Pretty and Rachel Hine [7] have collected data from roughly 9 million farmers in 52 countries, showing how yields can be increased, sometimes dramatically, using sustainable farming methods. These are:

1   Returning organic matter (manure, compost) to the soil.

2   Rotating crops, that is, planting different crops in rotation in any one field.

3   Including clover, mucuna bean, or other legume, in the rotation, which 'fixes' nitrogen, that is, takes nitrogen from the air and uses the energy of sunlight to convert it to nitrogenous compounds in the soil which fertilise the subsequent crop.

4   Using integrated pest management, that is, attracting beneficial insects which prey on the crop pest. For example, maize can be interplanted with grasses which attract the corn borer (a maize pest) and attract wasps which feed on the corn-borer larvae.

Taken together these four items constitute organic farming, but farmers do not need to take all four steps at once, if they are not seeking organic certification but primarily wish to maintain or improve yields, with reduced use of costly inorganic fertiliser and pesticides. Increasing the organic content of the soil helps water retention, reduces soil loss through erosion and, importantly, sequesters carbon (see section in Chapter 4) and so helps to combat global warming.

An interesting new development, now widespread in Brazil, is 'zero tillage'. The plough and hoe, symbols of farming these last 6000 years, are left to rust at the edge of the field. The mucuna bean or other legume, grown with or between harvested crops, fertilises the soil, suppresses weeds and forms a surface mulch. A rake (larger and heavier than a garden rake, of course, but light enough to be animal drawn) makes a row of furrows through the mulch into which the new seed is sown.

It is not claimed that these sustainable farming methods can increase yields to the level of, or to higher levels than, those of high-yield hybrid wheat, rice and maize on irrigated fields, with ample input of inorganic fertiliser and extensive use of pesticides. But many farmers in the South are farming rain-watered fields and cannot afford the use of inorganic fertiliser and pesticides. It is these farmers who benefit from integrated pest management and the use of legumes as natural fertiliser. The cost of inorganic fertiliser and pesticides is bound to rise, as the fossil fuels are phased out, or run out, so it is to these sustainable methods that we must turn, and to which the countries of the South must turn, to feed the world, in the long term, reaping the benefit of a clean environment and increased biodiversity.

SUMMARY AND CONCLUSIONS

References in this section, such as [11], refer to the Notes for Chapter 6.

At the time this last section is being written ('05) the Kyoto Protocol is coming into force. Ratification by Russia was a vital issue, since the terms of the protocol dictate that the treaty cannot be implemented until it has been ratified by a sufficient number of countries which, in total, contributed 55% of global emissions in 1990 [11]. Without both the US and Russia this condition could not be met. Emission reduction targets set by those countries which had ratified the treaty would then have become voluntary targets only, without penalty for failure to reach the declared aim.

From an economic viewpoint, the decision, for Russia, on whether to join or not was finely balanced. On the one hand, countries in Northern latitudes stand to gain from global warming with a longer growing season. Also, Russia depends heavily on oil and gas exports, so is interested to see rising demand for these fossil fuels. On the other hand, Russia could gain from carbon trading and St Petersburg is vulnerable to rising sea levels. From a political viewpoint, the decision was whether to side with Europe or with the US.

The targets set by the Kyoto Protocol, even if these are achieved by 2012, are only a start. By 2012, or sooner, if global warming is to be taken seriously, new negotiations must be started to set targets for greater reductions by 2020 and thereafter. If the climatologists are right, we can expect increasingly violent and extreme weather events to make clear the need for new international agreements. To be effective, these negotiations must include the US, Australia (also not yet a signatory) and also the developing nations, notably the populous countries of China and India. Perhaps the only basis on which agreement might be achieved would be some form of per capita allowance as outlined by Aubrey Meyer's curves (p.106). Emissions from aviation and marine fuels must be included, allocated perhaps to the country within which the plane or ship is fuelled.

Meanwhile, at some point, annual global production of conventional oil is going to peak and thereafter decline. This may happen as early as 2010 (projections from the Association for the Study of Peak Oil and Gas, reproduced on p.57). Once production can no longer meet potentially growing demand, the price of oil is bound to rise. To some extent, natural gas can substitute for oil, used directly, or as a source of hydrogen, but natural gas reserves may last little longer than oil reserves. Declining production of conventional oil and natural gas will stimulate development of renewable energy sources and, in some measure, achieve the reduction in fossil fuel use which is required to contol global warming. However, an international emission reduction treaty is still essential, to ensure that oil extraction from tar sands, and the continued use of coal, are allowable only if accompanied by carbon capture and safe storage (CCS). All this could be achieved through carbon trading which would reward those countries which develop renewable sources and CCS and penalise those countries which continue to burn fossil fuels heedlessly. Carbon trading would also reward those countries which conserve old forests or plant new permanent forests, while penalising forest destruction.

Aside from the development of the renewable sources of wind and PV, both the US and the EU may begin to set more agricultural land to wood biomass, to crops producing biofuels, and also to those crops, such as maize, from which plastics can be made (plastics today are made from fossil fuels). The US and the EU may no longer be producing surplus food for export.

In the scenario for the UK in 2050 set out in Chapter 5, it is supposed that coal will still be used at much the same level as today (with gasification, which traps all the heavy metals and other impurities in the coal in a stable slag, and with CCS). Renewable sources are developed to the maximum extent possible in this scenario, and total energy use is almost halved. This may serve as a good model for the global situation. Coal-burning power stations could be converted to gasification, with CCS, and coal could still be used at today's level, together with maximum development of renewable sources and some overall reduction in total energy use.

Both in the UK and globally, reduction in energy use will be driven by rising energy costs. The higher cost of oil will lead to increase in the cost of transport of all kinds, and hence to increasing cost for energy from other sources, since transport is so much involved in the manufacture and installation of wind turbines and PV panels, as well as in maintenance of the infrastructure required for coal mining and CCS. Increase in the price of oil will also increase the cost of fertilizer and pesticides. Food security, for developing countries, may increasingly depend on local production, rather than imports from the US and the EU, and on sustainable farming methods which return organic matter to the soil, rather than relying on artificial fertilizer, and rely on biological pest control rather than the application of pesticides. Recent experience, reported in this chapter, suggests that this may be the best way to feed a growing world population.

Of course, global trade will still be important, since each country must export to balance the import of goods which cannot be produced locally, and the less developed countries will still depend on the export of agricultural commodities rather than manufactured goods. As Peter Robbins shows, in his book *Stolen Fruit* [12], if the commodity-producing countries could get together, as the oil-producing countries have done, to manage supply, then the decline in commodity prices could be reversed, and this could free up land for much-needed food production. The increased cost of transport could lead to reduction in global trade with increase in trade at more local and regional levels.

I want to end with a quotation from Ivan Illich's *Energy and Equity* [13], leaving readers to decide whether or not they agree:

'Overconsumption of energy not only destroys the physical environment through pollution but, even more important, causes the disintegration of society itself. It is traffic, based on transport', he argues, 'which corrupts and enslaves, and results in a further decline of equity, leisure and autonomy for all'.

It is possible to see the declining availabilty of fossil fuels, and the need to reduce carbon dioxde emissions, as a favourable development in world affairs, leading to greater local and regional autonomy, perhaps some transfer of wealth from rich to poor countries to meet the emission reduction targets illustrated by the curves of p.106, and perhaps leading also to more sustainable and equitable food production and distribution.

We may well see global production of conventional oil peaking before 2020, forcing fundamental change in the global economy towards reduced energy use. A post-Kyoto international treaty may limit the use of coal unless accompanied by safe sequestration of carbon dioxide. If we take an optimistic view, these factors could allow stabilisation of atmospheric carbon dioxide at the 450 ppm set as a target on p.4 of this book.

The vital question, as the evidence becomes more compelling from day to day that global warming is gathering pace, is whether stabilisation at 450 ppm will be sufficient to prevent run-away, uncontrolled warming and catastrophic climate change. With atmospheric carbon dioxide levels rising to 380 ppm towards the end of 2005, we are already seeing steady annual reduction in the area of Arctic ice, permafrost melting in Siberia, more rapid flow of the Greenland glaciers and slowing of the Gulf Stream [14]. The changes explored in this book may have to be implemented on a much shorter time scale than I have allowed.

NOTES

Introduction

1 Land of the midnight sums, Fred Pearce, *New Scientist,* 25 Jan '03.

2 www.wmo.ch and WMO press release 2 July '03.

3 *Climate Change*, Cambridge University Press, 1990. This report was compiled by a panel of 270 meteorologists and oceanographers from 25 counties with a further 200 involved in peer review of the initial draft. There was thus, already by 1990, almost complete consensus among experts in the field that global warming was happening, was man-made and potentially catastrophic if uncontrolled. For more recent reports see www.ipcc.ch/ and for a summary of the Third Assessment Report (2001) see www.ipcc.ch/pub/un/syreng/spm.pdf.

4 When the Gulf Stream reaches the Arctic region, between Greenland and the Barents Sea, its waters become colder and denser and sink to the sea floor. A return current flows south on the ocean floor, warming and rising in equatorial regions to maintain the Gulf Stream flow. Global warming is melting the ice sheet over Greenland and causing more rain and snow to fall in Russia, increasing flow in the rivers that feed into the Barents Sea. This increase in fresh water, mixing with the incoming Gulf Stream in the Arctic region, could reduce the salinity and density of the current sufficiently to affect its fall to the ocean floor, and hence reduce the flow of the Gulf Stream or stop it altogether. Average temperatures in Britain might then fall by 5 to 10°C. This would be only a long-term fear but for the fact that measurements of the salinity in the seas south of Iceland show that this has fallen dramatically over the period 1970 to 2003 (World Meteorological Organisation, *World Climate News*, June '04). This process, potentially so fatal for Britain, has already begun and the Gulf Stream flow has weakened by 30% over the period 1992-2006 (*Guardian*, 1 Dec '05 and Fred Pearce, *New Scientist,* 3 Dec '05).

If run-away global warming, the ultimate disaster, led to melting of the Greenland and West Antarctic ice sheets, this would lead to a sea level rise of some 14 metres, and melting of the East

Antarctic a further 55 metres (see John Gribbin, *Hothouse Earth*, Bantam Press, 1990). The waters would rise above the height of Nelson's Column, at Trafalgar Square in London.

5 Royal Commission on Environmental Pollution, 22nd Report, Appendix E: www.rcep.org.uk/newenergy.htm and the Performance and Innovation Unit Energy Review: www.number10.gov.uk/su/energy/1.html.

6 see *Earthmatters* (the Friends of the Earth journal) Spring '05.

Chapter 1

1 UK Government White Paper, Our Energy Future, Feb '03. Available on www.dti.gov.uk/energy/whitepaper/index.shtml.

2 The energy units used in these Notes are the kilowatthour (kWh) and its multiples:

$$1000 \text{ kWh} = 1 \text{ MWh (megawatthour)}$$
$$1000 \text{ MWh} = 1 \text{ GWh (gigawatthour)}$$
$$1000 \text{ GWh} = 1 \text{ TWh (terawatthour)}$$

1 kWh (1000 watt-hours) is the energy consumed by a 100W light bulb left on for 10 hours. Home energy consumption is shown in kWh on electricity bills (and also in kWh on gas bills in the UK).

The 'watt' itself, and its multiples, are a measure of the *rate* at which energy is consumed or generated. The light bulb rated at 100W consumes 100Wh in an hour. A wind turbine rated at 2 MW will produce 2 MWh of energy each hour, when the wind is blowing well. On average over the year, offshore around Britain, wind turbines produce, at best, only some 38% of peak power (see Note 3 below). To calculate the energy produced in one year by a 2 MW turbine we have to multiply by 24 (hours in the day) and 365 (days in the year):

$$38/100 \text{ x } 2 \text{ x } 24 \text{ x } 365 = \text{about } 6700 \text{ MWh/year.}$$

Domestic electricity consumption for the year 2000 was 111.6 million MWh (DTI Energy Statistics, 2001) and the number of homes in the UK is close to 25 million (National Statistics, Social Trends, 2002) so average electricity consumption per home is about 4.5 MWh/year. A 2MW turbine will thus supply power to 6700/4.5 = about 1500 homes.

3 Total UK electricity consumption for the year 2000 was 329 million MWh (DTI Energy Statistics, 2001). A 3 MW turbine produces about 10,000 MWh per year (calculated as in Note 2 above). It follows that 16,000 3 MW turbines will produce 160 million MWh/year or 49% of electricity needs.

We should note here that the figure I am using in these calculations for the ratio of average to peak power for an offshore wind turbine (the 'load factor') of 38% is optimistic and allows for any possible technical improvement in wind turbine design. Other analysts assume 35% for offshore wind (see website reference in Note 4 for this chapter). The number of turbines needed to produce a given amount of power may thus prove to be slightly *larger* than the estimates given in this book, particularly if we also allow for energy loss in transmission. For onshore turbines in Britain the load factor is 25–30%.

The figures for final energy consumption for the year 2000 (DTI Energy Statistics, 2001) are:

	Million tonnes oil equivalent	TWh	%
Coal	4.4	51	3
Gas	60.0	698	37
Oil	66.5	773	41.5
Electricity	28.3	329	18
Renewables		9	0.5
		1860	

Transport (petrol, diesel fuel, aviation fuels) accounts for 72% of oil use.
For the latest figures see www.dti.gov.uk/energy/inform/dukes/index.shtml.

The period 2005–2004 shows a small increase in total energy use with increase in the use of oil (for transport) and increase in use of electricity, but also a significantly increased renewables contribution:

	2000 TWh	2004 Twh
Coal	51	36
Gas	698	652
Oil	773	811
Electricity	329	340
Renewables	9	33
	1860	1872

4 www.dti.gov.uk/energy/leg_and_reg/consents/future_offshore/index.shtml. The potential for offshore wind generation is estimated to be 697 TWh/year within territorial waters up to 30 m in depth (twice today's electricity consumption, see Note 3 above) and 3213 TWh/year in depths up to 50 m, if wind farms extend beyond territorial waters (more than today's total energy consumption).

5 Pure plant oil. Clean energy fuel, today and tomorrow, Niels Anso and Jacob Bugge, *Sustainable Energy News* No. 34, Aug. '01. To make biodiesel oil, vegetable oil is warmed to around 40°C and mixed in 5:1 proportion with 17.5 grammes/litre caustic soda (NaOH) in methanol. Glycerol settles as a sediment and the biodiesel can be decanted off and warmed to 70°C to drive off residual methanol. The rather thick (viscous) vegetable oil is converted to a thinner oil suited to diesel engines. With care, and with understanding of the toxic nature of methanol and caustic soda, this process can be carried out in a farm outbuilding.

6 The area of agricultural land in the UK, including grassland, is 18 million hectares (UK National Statistics, Social Trends, 2002). One hectare of oil seed rape will produce 1000 litres of oil per year (see *From the Fryer to the Fuel Tank,* Joshua and Kaia Tickell, pub. Green Teach, 1999). This is sufficient to run an average car for one year (10,000 miles at 45 miles per gallon). One tenth of UK agricultural land, 1.8 million hectares, can only fuel 1.8 million cars. There were over 24 million cars in Britain at the end of year 2000 (UK National Statistics, Social Trends, 2002) so 1.8 million hectares of oil seed rape could only provide fuel for 7.5% of today's cars.

Table of calorific values for different fuels

Seed oil	10 MWh/tonne
Mineral diesel oil	12 MWh/tonne
Coal	9 to 10 MWh/tonne
Wood (air dried)	3.8 MWh/tonne
Wood (oven dried)	5.3 MWh/tonne
Domestic waste	2 to 4 MWh/tonne
Natural gas	9.7 kWh/cubic metre
Hydrogen	2.8 kWh/cubic metre

For the gases, these values are per cubic metre at atmospheric pressure. Data from Gerald Foley, *The Energy Question,* Penguin, 1987 and Brian Horne, *Power Plants,* Centre for Alternative Technology, Powys, Wales, 1996.

7 Greenpeace have published a map (in their report 'Unlocking the Power of our Cities', '95) showing how much sunlight energy falls on the UK, on average over the year. This ranges from 2.9 kWh/sq.m./day on the south coast to 2.3 kWh/sq.m./day in the north of Scotland (these figures, of course, are based on records to date and do not tell us how they may change as a result of global warming).

At Oxford the energy from the sun is 2.7 kWh/sq.m./day and Susan Roaf has shown that the BP Solar panels achieve an efficiency of 10.8% (UK Department of Trade and Industry report, ETSU S/P2/00236/00/00). Her 30 sq.m. of PV thus generates, on average over the year

30 x 2.7 x 0.108 x 91 = 800 kWh/quarter (that is per 91 days)

We have to allow for loss at the inverter, bringing this down to about 700 kWh/quarter, of which Susan Roaf uses 200 kWh/quarter for the electric car and 500 kWh/quarter for domestic consumption. Although the Roaf family have a clothes washing machine, a dish washer, and two computers, they have managed to reduce their domestic electricity consumption to a little less than half the national average (see Note 2 for this chapter) by use of low energy light bulbs and energy-efficient appliances. Although 500 kWh/quarter can be sufficient for lighting, fridge, freezer, computer, TV and the motor drives of washing machines (if energy-efficient appliances are used) it will not, in general, be sufficient to contribute to cooking, water heating, or such space heating as even a well-insulated house may require in winter.

8 UK Department of Trade and Industry report, ETSU S 1365-P1, The Potential Generating Capacity of PV-clad Buildings in the UK. The amount of energy that could be generated from PV, on existing roofs and walls in the UK, is estimated at 162 TWh/year, if PV panels of 10% conversion efficiency are used, and 324 TWh/year if further research raises conversion efficiency to 20%. The authors of the report consider this level of efficiency could be achieved by 2020. Even on cloudy days, or on north-facing roofs and walls, diffuse daylight can contribute significant PV output.

9 Figures from Micropower, Seth Dunn, Worldwatch Paper 151, 2000, pub. Worldwatch Institute and from Greenpeace Business, Feb/March 2001.

10 Paul Brown, *Guardian*, 10 Feb '03.

11 Rooftop wind turbines of 1 kW generating capacity and 25% load factor would generate

25/100 x 1 x 24 x 91 = 546 kWh / quarter

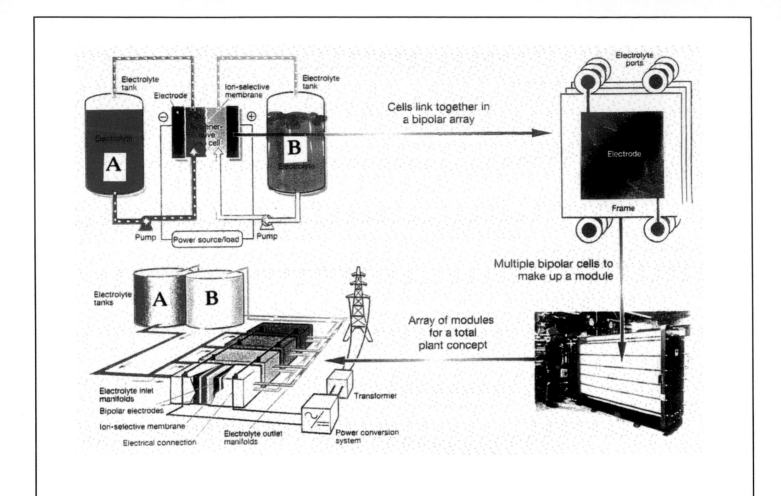

REGENESYS ENERGY STORAGE AT LITTLE BARFORD

Two tanks (A and B in the top left in this illustration) contain sodium bromide, in the one, and sodium polysulphide in the other, separated by a membrane permeable to sodium, but not to the bromide or polysulphide. When electricity is stored, sodium passes one way across the membrane. When electricity is needed, sodium moves in the reverse direction. The right side of the illustration shows that what is here described as a 'membrane' is really a stack of membranes connected in parallel, to produce, in total, a large area separating the two fluids.

(Source: RWE npower)

about half today's average electricity consumption per household.

12 *The Ecologist,* July/Aug '03. A lagoon of this type under construction off Swansea will supply half the city's electricity needs.

Chapter 2

1 The UK Government White Paper, Our Energy Future, Feb '03, gives an estimate of 26% for space heating, as a proportion of total energy use.

2 The surface of the sun is at about 5800°C and the radiation from the sun falls in the visible and near-infrared part of the spectrum, wavelengths to which glass is transparent. The interior of a glasshouse, or conservatory, is at a temperature of 10 to 300°C and radiation from objects at this much lower temperature is of longer wavelength, further into the infrared, and for these wavelengths glass is more opaque. Pilkington K and similar glasses used in modern double-glazing are coated, on one pane, with a transparent layer, which further reduces heat lost by radiation. Exactly how this is done is Pilkington's commercial secret, but one can imagine that, for the peak wavelength of radiation from a surface at 20°C, reflections from the glass-to-coating and coating-to-air surfaces are in phase, while transmissions from these two surfaces are out of phase.

3 For detail of the Millennium Green development see www.gustohomes.com. The simplest solar water heater is a black tank on the roof. This works well in sunny countries. In Britain, for efficient water heating, it is necessary to spend a little more on the installation and set up a row of vacuum tubes of total area about one square metre. Within each tube, which is about 100mm. in diameter and 1m. in length, there is a black strip, running the length of the tube, which absorbs the energy of sunlight. The heat is retained, because this strip is in a vacuum, and circulating fluid (water + antifreeze) carries the heat to the domestic hot water tank.

4 www.whispergen.com/whispgn.html. The Stirling engine (invented by Robert Stirling in 1816) converts heat to mechanical movement which, in the WhisperGen, drives an electricity generator. Surplus electricity can be fed into the National Grid with appropriate credit on electricity bills. The WhisperGen unit is marketed in the UK by Powergen.

5 Fred Pearce, *New Scientist* 13 July '02. In the production of today's concrete, a lot of energy is needed to heat the kilns to 1450°C. Additionally, a tonne of carbon dioxide is released into the atmosphere (from calcium carbonate) for each tonne of cement made. Cement manufacture is responsible for some 7% of total global carbon dioxide emissions. Cement can be made from magnesium carbonate (magnesite) at a lower temperature (650°C) and the magnesite concrete gradually reabsorbs the carbon dioxide released during its production.

6 R.H. Parker, *Proceedings of the Ussher Society*, Vol.7, pp. 316–320 (1991). A third borehole was drilled, as the research developed, to a depth of 2.6 Km, but to simplify the text in Chapter 2, I have described only the earlier work.

7 RE View, pub. UK Department of Energy, Spring '91.

8 When ice melts heat is absorbed. The water has undergone a phase change from solid to liquid. Coverings for internal walls are now available which contain capsules of material absorbing heat around 20°C (and changing phase) giving out this stored heat when the room temperature falls (John Shore in *The Green Building Bible,* 2nd ed, Green Building Press, 2005).

Chapter 3

1 *Factor Four,* Ernst von Weizsacker, Amory and Hunter Lovins, Earthscan, 1997.

2 K. Blok, R.H. Williams, R.E. Katofsky and C.A. Hendricks, *Energy*, Vol. 22, pp.161–168, 1997.

3 J.M.Ogden and J.Nitsch in *Renewable Energy*, eds. T.B.Johansson *et al.*, Island Press,1993.

4 120,000 3MW turbines could supply 1200 TWh/year (assuming that they produce on average 38% of peak power, see Note 3, Chapter 1). There are energy losses in electrolysis and in transmission so that the energy available, as hydrogen, from this number of turbines would be less than 1020 TWh/year, that is, less than 55% of total UK energy consumption (see Note 3, Chapter 1).

5 The average insolation in the Sahara is 300 W per square metre (*The Pherologist*, Vol. 4, No. 3, p.4, 2001). Allowing PV panels an efficiency of 14% in converting sunlight energy to electricity, the electrical power which could be generated from one square km of PV works out at:

$$14/100 \times 300 \times \text{one million sq.m.} \times 24 \text{ hours} \times 365 \text{ days} = 0.368 \text{ TWh/year.}$$

3000 sq.km of PV would produce 3000 x 0.368 = 1100 TWh/year. Allowing for energy loss in the electrolysis process and in transmission, the energy available, as hydrogen, from this area of PV would be less than 1020 TWh/year, that is, less than 55% of total energy use in the UK (see Note 3, Chapter 1).

6 *Fuelling Road Transport*, Nick Eyre, Malcolm Fergusson and Richard Mills, Energy Saving Trust (21 Dartmouth St, London SW1H 9BP), 2003, produced in collaboration with the Institute for European Environmental Policy and the National Society for Clean Air and Environmental Protection.

The current yield from short-rotation willow coppice is 10 oven-dried tonnes (odt)/hectare/year (see *Power Plants*, Brian Horne, Centre for Alternative Technology, Powys, Wales, 1996). Both Horne and Nick Eyre and co-authors, believe this yield can be increased to 20 odt/hectare/year in the future by use of improved strains of willow.

The calorific value of oven-dried wood chips is 5.3 MWh/odt (see Note 6 for Chapter 1). An area of 4 million hectares of willow coppice, then, could yield

$$4 \times 20 \times 5.3 = 424 \text{ TWh/year.}$$

This is calorific (heat) value and hydrogen can only be produced from wood chips with an efficiency of 65%, that is, retaining in the hydrogen 65% of the calorific value of the wood chips. The energy available in the hydrogen, derived from wood chips produced from 4 million hectares is then

$$65/100 \times 424 = 276 \text{ TWh/year.}$$

The energy used for transport in the UK today is 556 TWh/year (72% of 773 TWh, see Note 3, Chapter 1). If hydrogen-powered, fuel-cell cars and lorries are twice as efficient as today's, which are driven by internal combustion engines, in converting fuel energy into propulsion (as they are expected to be) then 4 million hectares of willow coppice could fuel future road transport at today's level of road use.

Nick Eyre and co-authors point out that this is a greater contribution to transport needs than crops grown for seed oil or ethanol could produce from this area and that, while rape or sugar beet require good arable land, willow coppice can grow well on poorer soils and in wetter and cooler conditions. It is suggested in their report that it might be a good idea to convert the wood chips to methanol (wood alcohol), a process which can be carried out with around 65% efficiency, at sites near the coppice areas, since methanol can be more readily transported than hydrogen to filling stations. At the filling stations, simple reforming plants could generate hydrogen from the methanol (see box on pp.136–137).

HYDROGEN OR METHANOL FROM WOOD CHIPS

The main components of wood (cellulose, hemi-cellulose) are carbohydrates with the general formula $[C.H_2O]n$. This is not surprising, since plants and trees in their magical variety, are made principally of carbon taken from the air and water taken from the soil.

At an experimental gasification station at the Battelle Columbus Laboratory in Ohio, USA, an input of 1650 odt/day of wood chips (odt = oven dried tonnes) is converted to 1.6 million cu.m./day of hydrogen (the hydrogen is produced under pressure, but 1.6 million is the volume in cubic metres which this gas would occupy at atmospheric pressure). Taking the calorific value of wood at 5.3 MWh/odt (see Note 4 for Chapter 2) this is an input of 8745 MWh/day. The calorific value of hydrogen is 2.8 kWh/cu.m. (see Note 6 for Chapter 1) so the output is 4480 MWh/day. Thus hydrogen is produced with an efficiency of 50%. Nick Eyre and co-authors (see Note 6 for Chapter 3) take conversion efficiencies for methanol or hydrogen production from wood chips as typically 65% today (see E.D.Larson, Technology for electricity and fuels from biomass, *Annual Review of Energy and the Environment* **18**:567–630, 1993). This is the figure used for calculations in Notes for Chapter 3.

The gas mixture after primary gasification is

H_2 31%

CO 41%

CO_2 11%

CH_4 14%

other gases 3%

We have seen (p.54) that methane (CH_4) can be reformed to hydrogen

$$CH_4 + 2H_2O = CO_2 + 4H_2$$

Hydrogen can also be derived from the conversion of carbon monoxide (CO) to carbon dioxide (CO_2)

$$CO + H_2O = CO_2 + H_2$$

By passing the gas from the primary gasification across appropriate catalysts, the gas mixture can be converted to 70% hydrogen, 30% carbon dioxide and small amounts of other gases.

If the aim is to produce methanol (wood alcohol) rather than hydrogen, the methane is first converted to hydrogen, as described above, then the gas stream is passed across a catalyst of zinc and chromium oxides at 350–400°C. This converts carbon monoxide and hydrogen to methanol

$$CO + 2H_2 = CH_3.OH$$

In a separate reaction, excess hydrogen reacts with CO_2 to produce more methanol

$$CO_2 + 3H_2 = CH_3.OH + H_2O$$

By these means, some 65% of the energy in the wood can be recovered in the form of methanol, readily transported from the gasification plants to service stations. At the service stations, hydrogen can be simply reformed from methanol

$$CH_3.OH + H_2O = CO_2 + 3H_2$$

(Information from Chapters 21 and 22 in *Renewable Energy*, T.B. Johansson and coauthors, Island Press. 1993).

7 The sustainable annual yields from UK woodlands range from 4 cu.m./hectare to 12 cu.m./hectare for beech, larch, ash and birch, and from 6 cu.m./hectare to 24 cu.m./hectare for pine, spruce, fir and cedar (see Forestry Commission, Booklet 48, *Yield Models for Forest Management*). If there is not too much change in the climate, or if Britain becomes wetter and warmer, an average yield for broadleaf woodlands might be around 8 cu.m./hectare and for conifers around 16 cu.m./hectare. The yield for mixed broadleaf and conifer would then average 12 cu.m./hectare, that is 6 oven-dried-tonnes (odt)/year. If we assume that 2.5 odt of this yield is available as sawn timber and 3.5 odt as wood chips (from forest clearance, from the branches of felled trees unsuited to the sawmill, and from sawmill waste) the energy content of the chips from 8 million hectares of woodland (calculated as in Note 6 for this chapter) would be

$$8 \times 3.5 \times 5.3 = 150 \text{ TWh/year.}$$

The energy content of the hydrogen which could be produced from these chips (again following the calculation of Note 6) would be

$$65/100 \times 150 = 100 \text{ TWh/year.}$$

In the scenario for the UK in 2050, in Chapter 5, it is supposed that an additional 0.5 million hectares has been set to willow coppice which, together with the yield from tree surgery and pruning in parks, gardens and orchards contributes, as hydrogen, a further 30 TWh/year.

The total area of the UK is 24 million hectares, made up of

crops and fallow	4.8 million hectares
grass and rough grazing	12.5 million hectares
forest and woodland	2.8 million hectares
urban and other	3.9 million hectares

(National Statistics, *Social Trends*, 2002).

The proposed increase in woodland, and the area set to coppice, would be on 'rough grazing', or poor quality, agricultural land.

8 To make a very rough estimate of how much dry waste wood might be available, let us assume that, in a sustainable society, new sawn timber is replacing old, in housing renewal. Taking figures from Note 7 for this chapter, 8 million hectares of woodland would be producing

$$8 \text{ million} \times 2.5 \text{ odt} = 20 \text{ million tonnes of sawn wood/year}$$

with a calorific (heat) value of

$$20 \times 5.3 = 100 \text{ TWh/year.}$$

CHP plants could use this wood to produce 30 TWh/year of electrical power and 60 TWh/year for heating.

Chapter 4

1 The theoretical output would have been

$$276 \text{ MW} \times 24 \text{ hours} \times 365 \text{ days} \times 27 \text{ years} = 65 \text{ TWh}$$

(1 TWh = one million MWh). In fact these reactors produced 40 TWh over their lifetime. This implies they were shut down for maintenance for about one third of the time. These were early reactors. Later reactors may require less downtime.

2 A 1000 MW nuclear power station produces, over its 50 year lifetime, a total of:

$$1000 \times 24 \text{ hours} \times 365 \text{ days} \times 50 \text{ years} = 438 \text{ TWh}$$
(1 TWh = one million MWh)

If we assume the station has to close down sometimes for maintenance, we can round this off to 400 TWh.

3 Jan Willem Storm van Leeuwen and Philip Smith (www.oprit.rug.nl/deenen/Technical.html) estimate that a nuclear power station must run for 10 years before there is net energy production, i.e. it takes 10 years to generate the energy used in ore extraction, construction of the station, running, decommissioning and eventual disposal of the waste. However, if uranium enrichment is achieved by centrifugation (with uranium in the gaseous form, as uranium hexafluoride) rather than by the earlier diffusion method, the energy used is less and the pay-back time is reduced to 3.5 years (see Ian Hore-Lacy in *Before the Wells Run Dry*, ed. Richard Douthwaite, Green Books, 2003). This payback time will become longer once supplies of higher grade ores are exhausted.

For any form of energy generation, calculation must always be made of net energy gain. Wind turbines are favourable, from this point of view. Running for 7 months they pay back the energy used in their construction, installation, servicing and eventual disposal (see Life cycle assessment of

Vestas 3MW offshore turbines, www.vestas.com).

PV panels are estimated to produce the energy used in their in their manufacture in the first five to nine years of their life (Wolfgang Sachs *et al.*, *Greening the North*, Zed Books, 1998). For their ultimate deployment in large numbers, as a sustainable energy source, it is essential that they be produced cheaply, and also with less energy used in their manufacture.

4 Bernt Kuemmel, *A Global Clean Fossil Scenario*, Roskilde University, Denmark, 1997.

5 A risk too far, Nicola Jones, *New Scientist*, 20 Oct. 2001. If storing carbon dioxide in the ocean reduced the pH of the water by only 0.1, this increase in acidity could be enough to harm sea life.

6 *The Hydrogen Economy*, Jeremy Rifkin, Polity Press and Blackwell, 2002.

7 14 million tonnes of domestic waste (calorific value 3 MWh/tonne, see Note 6 for Chapter 1) have a total calorific value of 42 TWh. One third of this can be converted to electricity (14 TWh/year). This is 4% of total electricity use (329 TWh/year, see Note 3 for Chapter 1).

8 UK Government White Paper, *Our Energy Future*, Feb '03.

9 www.renewableenergyaccess.com/rea/news/story?id=34133.

Chapter 5

1 *Stopping the Great Food Swap,* Caroline Lucas, The Greens / European Free Alliance in the European Parliament, 2001.

2 Figures for 1996, from The Perfect Pinta? SAFE, 1998. SAFE has now become Sustain (see list of companies and organizations).

3 *The Food Miles Report,* Angela Paxton, SAFE, 1994. SAFE has now become Sustain (see list of companies and organizations).

4 *Eating Oil,* Andy Jones, Sustain and Elm Farm Research Centre, 2001. The commodity category, in UK Transport Statistics, responsible for about one third of road transport is actually 'food, drink and tobacco'. To reduce transport it may be necessary to rely more on local beers, ciders and wines.

THE COST OF NUCLEAR-GENERATED ELECTRICITY

Cost of building a 1000 MW nuclear power station (1 MW = 1000 KW) £ ......
Cost of running this power station for 50 years (making fuel rods by mining uranium or
  reprocessing plutonium, replacing as necessary, disposal or reprocessing of spent fuel) £ ......
Cost of partial decommissioning (50 years after initial build) £ ......
Cost of final decommissioning to clear site (150 years after initial build) £ ......
Cost of ultimate storage of longlife radioactive components to be safe for 100,000 years £ ......
Cost of insurance against accident at the station, or in transport and storage of fuel rods £ ......

             Total cost £ A

For forward-projected costs we can put today's costs since, if nuclear power is to play an ongoing role, the building of a new power station must be accompanied by the decommissioning of an earlier one. To the total £ A we have to add the accrued interest which will have to be paid if we borrow £ A now, and pay it back from electricity sales over the next 50 years.

  £ A + interest = £ B.

Energy needed to build a 1000 MW power staion ....TWh
Energy needed to make fuel elements, mining uranium, etc., replace as necessary over
  50 years and dispose of, or reprocess, spent fuel ....TWh
Energy needed for partial and ultimate complete decommissioning ....TWh
Energy needed for ultimate storage of longlife radioactive material ....TWh

          Total energy used C TWh

Quantified list of radioactive substances discharged into atmosphere and seas ....

Electricity generated during the 50 year's life of the station (see Note 2, Chap. 4) 400 TWh

For the cost £ B, the net energy gain is 400 minus C TWh.

It is difficult to estimate the cost of ultimate disposal of long-life radioactive material, or possible long-term health costs arising from the release of radioactive material into air and seas. This box illustrates only the kind of calculation that has to be made to estimate the true cost of nuclear power.

(See also references in Note 3 for Chapter 4.)

5 *Farming and Food, a Sustainable Future,* Report of the Policy Commission on the Future of Farming and Food, chairman Sir Donald Curry, 2002.

6 *Regenerating Agriculture*, Jules Pretty, Joseph Henry Press, 1995.

7 Ecotaxes in Sweden, Germany and the Netherlands, see www.earth-policy.org/Updates/Updates14.htm.

8 www.evuk.co.uk

9 Taking electricity, transport and space heating as together responsible for 75% of total energy use, this leaves 25%, classified here as 'industry and services' (additional to electricity and space heating). 'Industry' here includes the year-on-year replacement of cars, domestic appliances, planes, trucks, railway rolling stock, renewal of housing stock, offices, shops, schools and hospitals, as well as day-to-day manufacture of concrete, plastics, fertilizer and chemicals, and the recycling of steel and paper. 'Services' includes the maintenance of roads, rail track and airports, as well as the maintenance of reservoirs, pipes and transmission lines for water, gas and electicity distribution.

10 www.gad.gov.uk

11 It is interesting to compare the 2050 scenario developed in this book with the 2050 'scenario 1' in the Royal Commission on Environmental Pollution (RCEP) 22nd report, Appendix E.

	RCEP scenario 1 (TWh/year)	My scenario (TWh/year)
onshore wind	57	25
offshore wind	100	125
PV	88	80
wave and tide	54	80
energy crops, agricultural and forestry waste, sawn wood waste	140	300
coal with CCS	455	330
oil and natural gas	930	0
	1824	940

My scenario places fewer wind turbines onshore, with correspondingly more offshore, and sources more energy from woody biomass, with correspondingly less from coal with CCS. There is then the important difference that the RCEP scenario takes total energy use to be much the same as today, with half this energy coming still from oil and natural gas. By contrast, I have supposed that the UK will have succeeded, by 2050, in reducing total energy use to half today's level. The RCEP scenario allows that nuclear power could substitute for the contribution allocated to coal with CCS.

12 Dry wood waste is estimated to contribute 30 TWh as electricity and 60 TWh as heat (see Note 8, Chapter 3) and municipal and farm waste are estimated to contribute 20 TWh as electricity and 40 TWh as heat (see Chapter 4).

Chapter 6

1 *Collision Course*, Andrew Simms, Ritu Kumar and Nick Robbins, New Economics Foundation, 2001.

2 *Stormy Weather, 101 Solutions to Global Climate Change,* Guy Dauncey and Patrick Mazza, 2001, New Society Publishers (www.newsociety.com).

3 If we take a mid-range figure of 2.5 trillion barrels for the amount of conventional oil originally available (see p.52) with 1 trillion used already, this leaves 1.5 trillion barrels, that is, 195 billion tonnes still available. If all this is used up, this will release about 165 billion tonnes of carbon. Using up the remaining conventional gas reserves will release a comparable amount of carbon. The allowable emission to stabilise the atmospheric carbon dioxide level at 450 ppm is 640 billion tonnes of carbon over the period 1990–2100 (Duncan Brack, Michael Grubb and Craig Windram, *International Trade and Climate Change Policies*, Earthscan, 2000).

4 *Escaping the Poverty Trap, The Least Developed Countries Report,* United Nations Conference on Trade and Development, 2002.

5 *When Corporations Rule the World,* David Korten, 2nd ed. Kumarian Press, 2001.

6 Lester Brown, *State of the World 2001,* Worldwatch Institute, Earthscan.

7 *Reducing Poverty with Sustainable Agriculture,* Jules Pretty and Rachel Hine, Centre for Environment and Society, University of Essex, 2001.

8 Peter Rosset, *Ecologist*, vol. 29, pp. 452–456, Dec 1999.

9 *The Promise of Sustainable Agriculture in Asia,* Jules Pretty and Rachel Hine, Natural Resources Forum, vol. 24, pp. 107–121, 2000.

10 *Agricultural Influences on Carbon Emissions and Sequestration,* Jules Pretty and Andrew Ball, Centre for Environment and Society, University of Essex, 2001.

11 Kyoto Protocol, see www.unfccc.int/

12 *Stolen Fruit,* Peter Robbins, Zed Books, 2003.

13 *Energy and Equity,* Ivan Illich, Marion Boyars, 1974.

14 Melting of polar sea ice, *Independent*, 16 Sept '05; melting of Greenland ice cap, *Independent on Sunday*, 20 Nov '05; melting of Siberian permafrost, *Guardian*, 11 Aug '05; slowing of Gulf Stream, *Guardian*, 1 Dec '05.

15 GNP/capita and Debt/capita for 2004 (Source: World Development Report, 2006)

	GNP/capita (US $)	Dept/capita (US $)		GNP/capita (US $)	Debt/capita (US $)
Bangladesh	440	134	Mozambique	250	258
Bolivia	960	631	Nigeria	390	252
Brazil	3090	1317	Peru	2360	1086
Burkina Faso	360	149	Philippines	1170	755
China	1290	149	Russian Fed	3410	1227
Congo Dem Rep	120	204	Tanzania	330	205
Ethiopia	110	102	Uganda	270	176
France	30,090	–	UK	33,940	–
Germany	30,120	–	US	41,400	–
Jamaica	2900	2068	Vietnam	550	192
Japan	37,180	–	Zambia	450	612
Mali	360	263			

COMPANIES AND ORGANISATIONS

Akeler
7 Clifford St
London W1S 2WE

Ballard Power Systems
Vancouver, Canada

B9 Energy Biomass
Unit 22
Northland Rd Industrial Estate
Derry, Northern Ireland BT48 0LD

Bill Dunster Architects
21 Sandmartin Way
Wallington
Surrey SM6 7DF

BioRegional Development Group
BedZED Centre
24 Helios Rd
Wallington, Surrey SM6 7BZ

British Nuclear Fuels
Reactor Decommissioning
Risley, Warrington
Cheshire WA3 6AS

British Photovoltaic Association
National Energy Centre
Davy Ave, Knowlhill
Milton Keynes MK5 8NG

British Wind Energy Association
1 Aztec Row
Berners Rd
London N1 0PW

Building Research Establishment (BRE)
Garston
Watford
WD25 9XX

Compact Power
Yara House
St Andrew's Rd, Avonmouth
Bristol BS11 9HZ

DaimlerChrysler UK
Tongwell
Milton Keynes MK15 8BA

Earth Energy
Falmouth Business Park
Bickland Water Rd
Falmouth TR11 4SZ

Energy Saving Trust
21 Dartmouth St
London SW1H 9BP
www.est.org/housingbuildings
Building a sustainable future
0845 129 7799
email: bestpractice@est.org.uk

Energy Technology Support Unit (ETSU)
ETSU reports are now lodged at the
British Library (Boston Spa)

Greenpeace
Canonbury Villas
London N1 2PN

Gusto Construction
(see under Millennium Green)

Holsworthy Biogas
Holsworthy
Devon EX22 7HH

Institute of Science in Society
29 Tytherton Road
London N19 4PZ

Marine Current Turbines
The Court, The Green
Stoke Gifford
Bristol BS34 8PD

Millennium Green
(housing development)
Builders: Gusto Construction
Millennium Green Business Centre
Rio Drive, Collingham
Nottinghamshire NG23 7NB

Nick Meers Photography
nick@meersphoto.com

Nigel Francis Photography
50 Hampden Drive
Kidlington
Oxfordshire OX5 2LS

Npower Renewables
Reading Bridge House
Reading Bridge
Reading, Berkshire RG1 8LS

Peabody Trust
45 Westminster Bridge Rd
London SE1 7JB

Progressive Energy
Swan House
Bonds Mill
Stonehouse
Gloucestershire GL10 3RF

Renewable Energy Association
17 Waterloo Place
London SW1Y 4AR

Rockwool
Pencoed
Bridgend CF35 6NY

RWE npower
Trigonos
Windmill Hill Business Park
Whitehill Way
Swindon, Wiltshire SN5 6PB

SAFE (see under Sustain)

Shell International
Shell Centre
London SE1 7NA

Soil Association
4056 Victoria St
Bristol BS1 6BY

Solar Century
9194 Lower Marsh
London SE1 7AB

Sustain(the Alliance for Better Food and Farming)
94 White Lion St
London N1 9PF

United Technologies Company
UTC Fuel Cells
195 Governor's Highway
South Windsor
CT 06074, USA

VentAxia
Fleming Way
Crawley, West Sussex RH10 2BR

Volvo UK
Globe Park
Marlow, Buckinghamshire SL7 1YQ

Welsh Biofuels
32 Chilcott Ave
Brynmenyn
Bridgend CF32 9RQ

Windelectric Management
Delabole
Cornwall PL33 9BZ

Woodfall Wild Images
17 Bull Lane
Denbigh
Denbighshire LL16 3SN

GLOSSARY

A.C.	alternating current
billion	one thousand million
biodiesel	oil made from seed oil suited to diesel engines
biodigestion	a liquid slurry of organic material in a closed tank (that is, in the absence of air) will be digested by bacteria in the slurry to produce biogas (composed mainly of methane)
biofuels	liquid fuels derived from plant material
biogas	the gas produced by biodigestion
biomass	organic material from which energy can be derived: seed oils, alcohols, wood chips, straw, animal slurry, sewage sludge, farm waste of all kinds
bioplastics	plastics made from plant material
calorific value	the amount of energy that can be extracted as *heat* (from a given mass of wood, coal, etc.). The amount of electricity or hydrogen which can be extracted will be less than this
catalyst	a metal which facilitates a chemical reaction (while itself remaining unchanged)
CCS	carbon capture and storage
CHP	combined heat and power, the cogeneration of heat and electricity
D.C.	direct current

district heating	a scheme, such as the one described on p.34 which distributes hot water through thermally insulated pipes, from a CHP plant to public buildings, or to houses on an estate
DTI	UK Department of Trade and Industry
ethanol	drinking alcohol, chemical formula $CH_3.CH_2.OH$
EU	European Union
gasification	when wood, coal or organic waste of low moisture content are heated in a stream of air they burn, and their stored energy is released as heat. When these combustible materials are heated in a closed tank in the absence of air (or in a restricted air flow) they are converted to a mixture of gases (hydrogen, natural gas, carbon dioxide, carbon monoxide, etc.). Most of the stored energy in the solid fuel is still present in the gas, which can thus be used for heating or electricity generation
GW	Gigawatt = 1000 Megawatts
GWh	Gigawatthour = 1000 Megawatthours
hectare	measure of area (100 metres by 100 metres)
kcal	kilocalorie (used as a measure of the energy in food). For health, the daily requirement is 2500–3000 kilocalories/capita/day, more for heavy manual work
kg	kilogramme
kW	kilowatt = 1000 watts
kWh	kilowatthour = 1000 watthours
£	pound sterling
LNG	liquified natural gas (natural gas become liquid at −162°C)

load factor	the ratio of average to peak power for a wind or tidal current turbine
LPG	liquified petrolium gas (mainly propane and butane)
m	metre
methane	the main component of natural gas (CH_4)
methanol	methyl alcohol, chemical formula $CH_3.OH$
mm	millimetre (one thousandth of a metre)
mph	miles per hour (50 mph = 80 kilometres per hour)
MW	megawatt = 1000 kilowatts
MWh	megawatthour = 1000 kilowatthours
NGL	natural gas liquids (liquid hydrocarbon by-products of natural gas extraction)
organic	'organic' has to be used, unfortunately, in two senses. One is for 'organic' vegetables, etc. produced without use of pesticides and satisfying other criteria determined by the Soil Association or similar bodies. The second sense is used when we talk of 'organic' material being returned to the land or digested to produce biogas. In this sense it means plant material, natural or digested, such as compost, animal slurry, straw, etc.
passive solar heating	the space heating derived from sunlight coming through south-facing windows and conservatories
ppm	parts per million
PV	photovoltaic (generating electricity from sunlight)
quarter	(as unit of time) period of three months
$	US dollar

short-rotation coppice	in the first year 1/3 of the area to be coppiced is planted with young stems. In the second year a further 1/3 is planted, etc. The stems are cut back after one year, to encourage strong growth, and harvested after three years of regrowth
sq.m.	square metre
sq.km.	square kilometre
tonne	one thousand kilogrammes
TW	Terawatt = 1000 Gigawatts
TWh	Terawatthour = 1000 Gigawatthours
UK	United Kingdom (England, Scotland, Wales and Northern Ireland)
US	United States of America
W	watt (a measure of the *rate* at which energy is generated or consumed)
Wh	watt-hour (a unit of energy, it is the energy produced in one hour by a source supplying energy at the rate of one watt). A 100W light bulb, alight for 10 hours, consumes one kWh
WTO	World Trade Organisation

MINOR SOURCES

These minor sources, mostly newspaper or magazine articles, are not as reliable as the main sources given in the Notes, but indicate general trends.

Introduction

The Gulf Stream keeps Western Europe 5–10°C warmer than it's latitude would otherwise permit, see Sir David King in the *Guardian*, 11 Mar '04.

Chapter 1

Environmental studies for North Hoyle offshore wind farm, see Environmental Statement, Innogy and National Wind Power, 2002.

Prediction of wetter weather in Britain and Northern Europe: Alan Thorpe, Director of the UK Meterological Office Hadley Centre for Climate Research, quoted in the *Guardian*, 13 Oct '00 and Tim Palmer of the European Centre for medium range weather forcasts, quoted in the *Guardian* 31 Jan '02.

Susan Roaf's house, *Times* 27 Jan '96, as well as ETSU report S/P2/00236/00/00.

Total area of the UK: 24 million hectares, UK National Statistics, Social Trends, 2002.

UK wind power plans: *Independent* and *Guardian*, 15 July '03.

Scroby Sands offshore wind farm, *Independent* 18 Apr '02.

Wind power in India, *Green Futures* May/June '00.

Coastline of UK 2800 miles, estimated from standard map.

Population of Denmark 5.4 million, *Independent*, 23 May '02.

Wind power from three States, North Dakota, Kansas and Texas sufficient for all US electricity need, Lester Brown, EcoEconomy, Earthscan, 2001.

2000 hectares of willow coppice would supply one third of wood chips needed for Eggborough ARBRE project, personal communication, First Renewables.

Detail of Buckfast Abbey small-scale hydro, personal communication, Bursar's Office.

PV in the third world, *Greenpeace Business,* Feb/Mar '01 and *Micropower,* Seth Dunn, Worldwatch Paper 151, July '00, Worldwatch Institute.

The roof tiles are called 'solar shingles', solarcentury.co.uk 0870 735 8100.

Prototype tidal current turbine at Lynmouth rated at 300 kW (later a 1 MW unit). A 40% load factor is expected, that is, average power related to peak power, comparable with offshore wind, personal communication, Jeremy Thake, IT Power.

Chapter 2

The U-value for normal double glazing is 2.8, the U-value for Pilkington K double glazing is 1.9, personal communication, Pilkington Glass.

Earth and rubber tyre homes, *Guardian* 4 June '03.

Drawing tidal energy slows the earth's spin, *The Energy Question*, Gerald Foley, Penguin, 1976.

Chapter 3

Average insolation (global) 240 W/sq. m., *Ecologist Report on Climate Change*, Nov '01.

The production of hydrogen from natural gas is about 86% efficient, compressing hydrogen for storage takes about 20% of its energy content, see reference given in Note 4 for Chapter 4.

Hydrogen-powered bike and scooter, *Green Futures* Jan/Feb '02.

Industrial development at Kalundborg, Denmark, *Green Futures* May/June '02.

EU legislation on end-of-life return of products, *State of the World '02,* Worldwatch Institute.

Internal combustion engine 20–25% efficient, fuel cells can be up to 80% efficient, *Green Futures* May/June '01.

Honda FCX, *Green Futures* Nov/Dec '02.

Peter Hoffman, *Tomorrow's Energy*, MIT Press, 2002, calculates that one million cars require 21 square miles of PV, that is 54 sq. km. There were 24 million cars in the UK in the year 2000 (see Note 6 for Chapter 1). This leads to an estimate of around 1300 sq. km of PV needed to supply hydrogen for UK car transport.

Ballard's contribution to fuel cell development: Peter Hoffman, *Tomorrow's Energy,* MIT Press, 2002.

Genetic engineering of bacteria for protein production, *New Scientist,* 3 Feb '96.

Chapter 4

Chernobyl and fallout in Cumbria region, *Ecologist*, Sept '01.

Cracks and corrosion in older nuclear reactors, see 'Are ageing US nuclear reactors safe?' *New Scientist*, 9 Aug '03.

A terrorist attack on Sellafield could release 100 times the radioactivity released at Chernobyl, making the North of England uninhabitable, *Guardian*, 10 Jan '02.

Sequestering carbon, New Scientist, 17 Mar and 31 Mar '01, *Scientific American,* Feb '00.

28 million tonnes of waste generated annually in UK, *Women's Environmental Network Newsletter,* Spring '01.

Tree planting in China, see *A Solar Manifesto*, Hermann Scheer, 2nd ed., pub. James and James, 2001 and *Natural Capitalism*, Paul Hawken and coauthors, Earthscan, 1999.

Using MOX as a fuel, more plutonium is created than is consumed (1.17 tonnes created for 1 tonne consumed) *Ecologist*, Feb '02.

Tar sands and oil shales, *The Energy Question*, Gerald Foley, Penguin, 1976.

Mature forests take up carbon, *New Scientist* 26 Oct '02.

Methane hydrates are trapped in ice, *New Scientist* 14 Dec '02.

11 million cows in the UK, Lester Brown, *EcoEconomics*, Earthscan, 2001.

Straw can be used as a fuel, but many farmers prefer to plough it back into the fields to improve the organic content and texture of the soil, Biomass as a Renewable Energy Source, Royal Commission on Environmental Pollution report, '04

Effect of CO_2 on pH of sea water, *Scientific American*, Feb '00.

Temperature at the centre of the sun: 15 million°C, Steven Weinberg, *The First Three Minutes,* Andre Deutsch, 1977.

Imported coal half cost of coal produced in the UK, *Guardian* 12 Feb '03.

Fred Pearce in *Global Warming*, Dorling Kindersley, 2002, gives storage capacity for carbon dioxide in the North Sea oil and gas fields as 15 Gigatonnes.

Fusion temperature 100 million°C, *Guardian*, 27 Nov '03.

Plankton blooms, see *The Blue Planet,* Andrew Byatt, Alastair Fothergill and Martha Holmes, BBC, 2001.

Chapter 5

Comparison of car and rail energy use, see Mark Walisiewicz, *Alternative Energy*, Dorling Kindersley, 2002, also Open University course material, *Travelling Light*.

Shopping and commuting by helicopter in Sao Paulo, *Guardian*, 7 Aug '00.

Chapter 6

Number of cars, worldwide, in 1992, 587 million, number in 2004, 855 million, *New Scientist Supplement*, Apr '01.

Increase in air travel, *Vital Signs '99–'00*, Worldwatch Institute.

1.1 billion without access to clean water, *Vital Signs '01–'02*, Worldwatch Institute.

Nearly half the global population live on less than $2 per day, *State of the World '01*, Worldwatch Institute.

60% of US food grown is exported, *Guardian Supplement*, 22 Aug '02.

GNP defined, Richard Douthwaite, *The Growth Illusion*, Green Books,1992.

GNP increase in '90s in the UK with reduced use of materials, *Independent*, 29 July '02.

The US has abundant desert and windy areas for generation of hydrogen by electrolysis, Peter Hoffman, *Tomorrow's Energy*, MIT Press, 2002.

30,000 native varieties of rice in India, see *Natural Capitalism*, Paul Hawken, Amory Lovins and Hunter Lovins, Earthscan, 1999.

The role of forests in carrying rain inland, see *EcoEconomy*, Lester Brown, Earthscan, 2001.

INDEX